Improving the Effectiveness of U.S. Climate Modeling

Panel on Improving the Effectiveness of U.S. Climate Modeling

Board on Atmospheric Sciences and Climate

Commission on Geosciences, Environment, and Resources

National Research Council

NATIONAL ACADEMY PRESS
Washington, D.C.

NATIONAL ACADEMY PRESS • 2101 Constitution Avenue, NW • Washington, DC 20418

NOTICE: The project that is the subject of this report was approved by the Governing Board of the National Research Council, whose members are drawn from the councils of the National Academy of Sciences, the National Academy of Engineering, and the Institute of Medicine. The members of the committee responsible for the report were chosen for their special competences and with regard for appropriate balance.

This study was supported by Contract No. 50-DKNA-6-90040 between the National Academies and the National Oceanic and Atmospheric Administration and Grant No. ATM-9814235 between the National Academies and the National Science Foundation. Additional support for this study was provided by the Department of Energy and the National Aeronautics and Space Administration. Any opinions, findings, conclusions, or recommendations expressed in this publication are those of the author(s) and do not necessarily reflect the views of the organizations, agencies, or subagencies that provided support for the project.

International Standard Book Number: 0309-07257-3
Library of Congress Control Number: 2001088954

Additional copies of this report are available from National Academy Press, 2101 Constitution Avenue, N.W., Lockbox 285, Washington, D.C. 20055; (800) 624-6242 or (202) 334-3313 (in the Washington metropolitan area); Internet http://www.nap.edu

THE NATIONAL ACADEMIES

National Academy of Sciences
National Academy of Engineering
Institute of Medicine
National Research Council

The **National Academy of Sciences** is a private, nonprofit, self-perpetuating society of distinguished scholars engaged in scientific and engineering research, dedicated to the furtherance of science and technology and to their use for the general welfare. Upon the authority of the charter granted to it by the Congress in 1863, the Academy has a mandate that requires it to advise the federal government on scientific and technical matters. Dr. Bruce M. Alberts is president of the National Academy of Sciences.

The **National Academy of Engineering** was established in 1964, under the charter of the National Academy of Sciences, as a parallel organization of outstanding engineers. It is autonomous in its administration and in the selection of its members, sharing with the National Academy of Sciences the responsibility for advising the federal government. The National Academy of Engineering also sponsors engineering programs aimed at meeting national needs, encourages education and research, and recognizes the superior achievements of engineers. Dr. William A. Wulf is president of the National Academy of Engineering.

The **Institute of Medicine** was established in 1970 by the National Academy of Sciences to secure the services of eminent members of appropriate professions in the examination of policy matters pertaining to the health of the public. The Institute acts under the responsibility given to the National Academy of Sciences by its congressional charter to be an adviser to the federal government and, upon its own initiative, to identify issues of medical care, research, and education. Dr. Kenneth I. Shine is president of the Institute of Medicine.

The **National Research Council** was organized by the National Academy of Sciences in 1916 to associate the broad community of science and technology with the Academy's purposes of furthering knowledge and advising the federal government. Functioning in accordance with general policies determined by the Academy, the Council has become the principal operating agency of both the National Academy of Sciences and the National Academy of Engineering in providing services to the government, the public, and the scientific and engineering communities. The Council is administered jointly by both Academies and the Institute of Medicine. Dr. Bruce M. Alberts and Dr. William A. Wulf are chairman and vice chairman, respectively, of the National Research Council.

PANEL ON IMPROVING THE EFFECTIVENESS OF U.S. CLIMATE MODELING

EDWARD S. SARACHIK (*Chair*), University of Washington, Seattle
LENNART BENGTSSON, Max-Planck-Institut für Meteorologie, Hamburg, Germany
MAURICE L. BLACKMON, National Center for Atmospheric Research, Boulder, Colorado
MARGARET A. LEMONE, National Center for Atmospheric Research, Boulder, Colorado
ROBERT C. MALONE, Los Alamos National Laboratory, Los Alamos, New Mexico
MATTHEW T. O'KEEFE, University of Minnesota, Minneapolis
RICHARD B. ROOD, NASA Goddard Space Flight Center, Greenbelt, Maryland
STEPHEN E. ZEBIAK, International Research Institute for Climate Prediction, Palisades, New York

STAFF
VAUGHAN C. TUREKIAN, Study Director
ALEXANDRA R. ISERN, Program Officer
CARTER W. FORD, Project Assistant

CLIMATE RESEARCH COMMITTEE

Preface

Information derived from climate modeling has become increasingly important in recent years. Seasonal-to-interannual forecasts of the global aspects of El Niño/Southern Oscillation (ENSO) have been made and have proven valuable in both public and private applications.

1. Patterns of global climate, especially the North American/Arctic Oscillation and the Pacific Decadal Oscillation, have been shown to strongly affect regional climate, raising questions about the mechanisms and the predictability of these patterns.

2. Long-term climate change and the response of the climate to the anthropogenic emissions of radiatively active gases and constituents have been intensively studied over the last thirty years, with the results being scrutinized to evaluate possible mitigation and adaptation.

3. Regional assessments of climate variability and change have begun and this has led to an increasing awareness of the intricate interactions of the physical climate, ecological systems, and human institutions.

More and more we understand that climate variability and change impacts society and that dealing with climate-related disasters, conflicts, and opportunities requires the best possible information about the past, present, and future of the climate system.

It is in this context that the National Research Council (NRC) report *Capacity of U.S. Climate Modeling to Support Climate Change Assessment Activities* (NRC, 1998a) pointed out that the United States now lags behind other nations in its ability to model the climate. At a time of increased need came a message of decreased capacity. This present report is a response, a *first* response, to that report.

To address the issues involved in improving this situation, the NRC empanelled the authors of this report and charged them with examining the computer and human resource issues involved in assessing U.S. climate modeling needs, especially at the high-end of modeling. The panel itself represented a wide range of expertise in climate and climate modeling, but to supplement its expertise a survey was conducted to gain an appreciation of the magnitude of the issues and to elicit opinions from the modeling community about our present plight and possible solutions. A general meeting of modelers was held at the National Academies on August 21, 2000, to further hear the concerns of the climate modeling community. Considering all these sources of input, the panel deliberated its recommendations and produced this report.

The panel recognizes that one of the most important inadequacies of this report is its inability to place climate modeling fully in the context of the panoply of issues arising from the interaction of physical climate, ecosystems, and human institutions The problem is just too big, as was illustrated in a previous NRC report (NRC, 1999a—'Pathways'). The panel hopes that this broader context will be recognized and will continue to be addressed in the future.

The discussion of computer architectures reflects the updated information available during panel deliberations and report preparation. Because the field of computer technology is fluid and rapidly evolving, upgrades in computing systems, such as NCEP's recent acquisition of an IBM Power-3 Winterhawk-II, which occurred after the preparation of the report, are not reflected in the report's summary data on computer performance (e.g., Table 3.1). The panel does not believe that such upgrades would change its overall findings or recommendations.

The panel would like to acknowledge the dedicated industry of Dr. Alexandra Isern, Dr. Vaughan Turekian, and Mr. Carter Ford, without whom the production of this report would have been impossible.

E. S. Sarachik
Chair

Acknowledgments

The panel greatly appreciates the efforts of all of the survey respondents whose input was critical for the completion of this report. Roberta Miller is greatly acknowledged for her comments and corrections on the surveys prior to their distribution. In addition, the panel would like to thank all of those who participated in the workshop and provided input to the discussions at this meeting. Bob Atlas is acknowledged for providing helpful contributions to this report and Tom Bettge for providing us with a figure. The panel would also like to acknowledge useful discussions with Andy White, Rod Oldehoeft, Tom Ackerman, Chris Davis, Dennis Joseph, Chin-Hoh Moeng, David Parsons, Chris Snyder, Wojtek Grabowski, Xiaoqing Wu, Jim Hoke, Bill McCracken, and Bruce Webster.

The authoring group would like to thank all of those who participated in the review and provided input to the discussions at the August 21-23, 2000, workshop. This report has been reviewed in draft form by individuals chosen for their diverse perspectives and technical expertise, in accordance with procedures approved by the NRC's Report Review Committee. The purpose of this independent review is to provide candid and critical comments that will assist the institution in making its published report as sound as possible and to ensure that the report meets institutional standards for objectivity, evidence, and responsiveness to the study charge. The review comments and draft manuscript remain confidential to protect the integrity of the deliberative process. We wish to thank the following individuals for their review of this report:

ERIC BARRON, Pennsylvania State University, University Park
ALAN BETTS, Atmospheric Research, Pittsford, Vermont
DAVID DENT, European Centre for Medium-range Weather Fore-
     casts, Reading,United Kingdom
INEZ FUNG, University of California, Berkeley
ANTHONY HOLLINGSWORTH, European Centre for Medium-
     range Weather Forecasts, Redding, United Kingdom
JERRY MAHLMAN, National Ocean and Atmospheric Administra-
     tion, Geophysical Fluid Dynamics Laboratory, Princeton, New
     Jersey
DANIEL SAREWITZ, Columbia University, New York, New York

Although the reviewers listed above have provided many construc-
tive comments and suggestions, they were not asked to endorse the conclusions or recommendations nor did they see the final draft of the report
before its release. The review of this report was overseen by Eugenia
Kalnay, University of Maryland, College Park, appointed by the Commission on Geosciences, Environment, and Resources, and Alexander Flax,
Potomac, Maryland, appointed by the NRC's Report Review Committee,
who were responsible for making certain that an independent examination of this report was carried out in accordance with institutional procedures and that all review comments were carefully considered. Responsibility for the final content of this report rests entirely with the authoring
committee and the institution.

Contents

Executive Summary

BACKGROUND AND CHARGE

The U.S. research community is a world leader in the study and understanding of climate and climate variability. Modeling capabilities are being applied to the study of anthropogenic impacts on the climate system, such as those resulting from additions of radiatively active constituents to the atmosphere, and from slow natural variations of climate. Short-term climatic variations, such as those occurring with the El Niño/Southern Oscillation, are becoming better understood and may be increasingly predictable as a result of the observations and modeling by this community.

This skill of predicting short-term climate variations and the information gained to better understand natural variability and the response to natural and anthropogenic perturbations is of great societal, ecological, and economic value for future planning. Recently a key use of climate models has been the production of legally mandated climate assessments (the U.S. National Assessment) and assessments required by international agreement (the assessments of long-term climate change performed by the Intergovernmental Panel on Climate Change and the Ozone Assessments called for by the Montreal protocols). Regional assessments to characterize climate impacts on a more local scale are increasingly in use, as they become valuable for planning purposes in both the public and private sectors.

Recognizing the societal importance of climate modeling, the National Oceanic and Atmospheric Administration (NOAA) and the National Science Foundation (NSF) requested the Climate Research Committee (CRC) of the National Research Council (NRC) to investigate the current state of

U.S. climate modeling and its ability to meet these assessment demands. In response to the request, the NRC produced a report entitled *Capacity of U.S. Climate Modeling to Support Climate Assessment Activities* (NRC, 1998a). This report evaluated allocation of resources to high-end modeling and whether these resources were being used effectively. The CRC found that "insufficient human and computational resources are being devoted to high-end, computer intensive, comprehensive modeling, perhaps, in part, because of the absence of a nationally coordinated modeling strategy." This present study focuses on the challenges posed in the 1998 report and as specified in the statement of task given to the panel.

The purpose of this study is to provide relevant federal agencies and the scientific community with an assessment of the nation's technical modeling needs and a vision of how government, interacting with the rest of the scientific community, can optimize the use of modeling talents in the United States. This study addresses the challenges posed in the Climate Research Committee's 1998 report, *Capacity of U.S. Climate Modeling to Support Climate Change Assessment Activities*. In pursuit of these objectives, the panel:

1. Examines the major types of climate modeling, paying particular attention to both the similarities (e.g., potential synergisms) and unique characteristics of each. Specific issues to be addressed include model construction and testing, data input and archival, ensemble simulation, interrogation and diagnostics, evaluation, and operational utilization.

2. Describes the computational and human resources required to effectively conduct climate modeling in the United States to meet the needs of the climate applications, policy, and scientific communities. This evaluation will include consideration of shifts in computational architectures and potential for, and cost of, improvements in model codes. It will also consider the utilization of common climate modeling tools, protocols, and data, and the availability of cooperative opportunities between different scales of modeling effort and institutions.

3. Quantitatively assesses the computational and human resources that are presently directed toward climate modeling in the United States.

4. Describes ways in which the efficacy of the U.S. climate modeling enterprise might be improved, given the current needs and resources. The report will define a set of issues that are fundamental to the enhancement and sustenance of climate modeling in the United States.

CLIMATE MODELS AND OBSERVATIONS

Climate models are mathematical representations of the major systems (atmosphere, ocean, land, snow, and ice) whose interactions determine climatic means and climate variability. For the most complex models, components of the climate system are linked, or coupled, using algorithms describing the connections between these systems. Due to limitations in resolution such components as radiation, clouds, and turbulent processes are generally unresolved in climate models and therefore require separate numerical representation. Although the climate components in these models can be separately built and evaluated, the nature of their coupling determines the behavior of the climate model. To evaluate model realism, model outputs are compared to each other and to environmental observations. The results of these comparisons form the basis for changes to model code, which improve the mathematical representation of physical processes.

Model simulations can be used for short-term environmental prediction, climate prediction, and assessment of future climatic responses to anthropogenic forcing. They can be combined with observations to produce model-assimilated data sets, and to design and improve climate observing systems. Atmospheric analysis through the assimilation of weather data into weather forecast models, and the need to downscale and interpret the output of climate data to the local region, provides a continual and necessary interaction between climate modeling and weather modeling. Because models are tested and improved through comparison to observational data, progress in modeling and observations are interdependent. An effective and integrated system for producing and delivering climate information needs to be supported by data collected from a dedicated climate observing system. Because the present atmospheric observing system was built primarily for weather prediction, and as such is subjected to major changes in time, it is inadequate to unambiguously detect and monitor climate change. Without regular and systematic analysis of parameters controlling the climate system, it is impossible to clearly document climatic variability and long-term climate trends. The panel therefore notes that the lack of a suitable sustained observing system for climate limits progress in climate modeling.

COMPUTING RESOURCES

The building of parameterizations of individual model elements; the running of uncoupled atmosphere, land, and ocean models; and the diagnoses and analyses of coupled climate model outputs can be accomplished at the workstation level. The integration of the components into comprehensive coupled climate models, the running of these coupled models, and the integration of global data with models can be attained only by using the very highest end of supercomputers.

The panel concludes that sustained computational capabilities of 10-

100 Tflops[1] would meet the needs of the different types of climate modeling; this capability is almost attainable using present technology. The potential for using this capability to achieve adequate throughput is determined by the efficiency of a given model code on the available supercomputer architecture (parallel vector processors and massively parallel commodity processors). The panel concludes that parallel vector computers provide superior processor speeds, greater usability, and lower human resource requirements; however, the massively parallel commodity processor machines are currently the only ones that can be purchased in the United States. The primary drawback to a massively parallel architecture is that the speed at which climate modeling code will run on a large number of parallel processors does not linearly increase with the number of processors but is controlled by Amdahl's law. This law essentially states that incomplete parallelization of model code creates significant computational inefficiencies and reduces the speed at which that code is run on a large number of processors. Even perfectly written code must deal with the irreducibly sequential underlying dynamics so that there is an absolute theoretical limit with massively parallel machines to the speedup factor possible over the speed of a single processor. This limit is far less than the theoretical maximum based on the number of processing elements.

As part of this study a survey was conducted to quantitatively assess the computational and human resources presently directed toward climate modeling in the United States. The survey responses indicated that access to increased computational power is desired across all modeling scales but is most apparent at the highest end. Smaller and intermediate-size modeling groups are able to accomplish modeling undreamed of a generation ago, but they expressed the desire for increased access to supercomputing facilities. A recurrent theme in the survey results was the difficulty of hiring and retaining computer technologists because of extraordinary competition from the information technology sector.

The survey also provided information on supercomputing capabilities for climate modeling in the United States. With the exception of a few centers devoted to prediction, most of the computational load in existing centers is devoted to modeling for research purposes. Computing capabilities at large modeling centers have sustained speeds between 10 and 100 Gflops (with most being toward the lower end of the range) on actual model codes using massively parallel processing systems. These systems enable present coupled climate models to be run for hundreds of model years at resolutions of 300 km in the atmosphere and approximately 100

---

[1]Computer speed is frequently measured in units of "floating operations per second", or flops. Megaflops (Mflops) indicates a speed of a million operations per second. Gigaflops (Gflops) equal 1000 Mflops, Teraflops (Tflops) equal 1000 Gflops.

km in the ocean. One of the major problems faced by large modeling centers is the conversion of existing model codes, previously optimized for machines using small numbers of vector processors, to those that can efficiently run on parallel architectures. The scarcity of human resources in information technology further compounds this problem.

RESEARCH AND OPERATIONS

During the panel's examination of measures to improve the effectiveness of U.S. climate modeling, a distinction between modeling in response to societal need ("operational" modeling) and modeling for research arose from the necessarily differing roles played by the research community in each. Societal requirements for the regular delivery of useful climate information products places demands on the research community that are difficult to meet because of insufficient resources, the lack of research organization capable of concentrating the resources needed to respond to these demands, and an inappropriate management structure to carry out the regular and systematic production of products. Although operational modeling depends on research for its success, by itself, it is not a research activity and cannot be well addressed in a research culture. The panel concluded that the present research infrastructure spread among many research agencies, each operating in its own interests according to its own culture, is not capable of responding to the modeling demands of regular assessment and prediction, nor is the management structure of the U.S. Global Change Research Program (USGCRP) able to instill such a culture or otherwise provide the focus required for regular climate product production.

Analogous to operational weather forecasting, centralized climate modeling activities and the maintenance of a climate observing system should exist outside the research domain but should have a close interaction with the research community. The beneficial interaction between research and operational climate modeling communities requires that different groups be able to run a variety of models, interchange model components and parameterizations, and compare output and observational data in common formats. This "Common Modeling Infrastructure" is defined as a set of standards, protocols, and associated tools for physical parameters, model codes, file formats, diagnostics, visualization, and data storage, as well as an 'exchange infrastructure' that fosters efficient collaboration among modeling groups.

The state of the climate system can only be defined using sustained observations of critical components of the climate system. Observational data assimilated into a comprehensive coupled climate model will enable the verification and enhancement of model code to produce accurate and continual climate analysis. From panel deliberations and acknowledgments of the robust linkages between research, observations, and climate modeling, the panel endorses previous NRC reports (NRC, 1998b, 1999a, 1999b, 1999c, 2000a, 2000b) that called for the development of a sustained climate observing system.

RECOMMENDATIONS

Based on panel expertise, input from a one-day workshop held as part of this study, and from a survey distributed to large, intermediate-size, and small climate modeling centers, the panel makes the following recommendations.

The Need for Centralized Operations

Recommendation 1: In order to augment and improve the effectiveness of the U.S. climate modeling effort so that it can respond to societal needs, the panel recommends that enhanced and stable resources be focused on dedicated and centralized operational activities capable of addressing each of the following societally important activities:

1. short-term climate prediction on scales of months to years;
2. study of climate variability and predictability on decadal-to-centennial time scales;
3. national and international assessments of anthropogenic climate change;
4. national and international ozone assessments;
5. assessment of the regional impacts of climatic change.

The Need for Open Access to the Most Appropriate Computer Architecture

Recommendation 2: The panel recommends the adoption of a scientific computing policy ensuring open access to systems best suited to the needs of the climate modeling community.

Recommendation 3: Researchers should have improved access to modern, high-end computing facilities connected with the centralized operational activities discussed in Recommendation 1. These facilities should be sufficiently capable to enable comprehensive study of the climate system and help develop models and techniques to address relevant high-end climate modeling problems.

The Need for a Common Modeling Infrastructure

Recommendation 4: In order to maximize the effectiveness of different operational climate modeling efforts, these efforts should be linked to each other and to the research community by a common modeling and data infrastructure. Furthermore, operational modeling should maintain links to the latest advances in computer science and information technology.

Human Resource Needs in Support of Climate Modeling Activities

The climate modeling community faces a severe shortage of qualified technical and scientific staff members, who, because of high salaries and incentives, find the high-tech industry more desirable than the research and operational modeling centers. (Some overseas groups, such as the European Centre for Medium-range Weather Forecasts (ECMWF), have overcome this difficulty by offering lucrative salary packages that U.S. modeling groups have been unable to match.) A further complication is the dependence of university-based modeling groups on the vagaries of short-term funding for employee salaries. The state of affairs also affects graduate programs, wherein universities see many students accepting attractive offers from private industry prior to completion of their degrees. This strain on human resources has resulted in declining graduate enrollments in all areas of the climate sciences and in the growing disparity in the quality of life of scientists—especially young ones—and their private sector counterparts.

The shortage of highly skilled technical workers is, however, not unique to the climate modeling community; it is part of a larger shortage affecting nearly all areas of science and engineering except those with strong linkages to the private sector. The complexity of the problem and the lack of expertise on the panel to address this issue precludes this panel from making any specific recommendations related to human resources.

Institutional Arrangements for Delivery of Climate Services

Recommendation 5: Research studies on the socio-economic aspects of climate and climate modeling should be undertaken at appropriate institutions to design the institutional and governmental structures required to provide effective climate services. This assessment should include:

1. an examination of present and future societal needs for climate information;

2. a diagnosis of existing institutional capabilities for providing climate services;

3. an analysis of institutional and governmental constraints for sustaining a climate observing system, modeling the climate system, com-

municating with the research community, and delivering useful climate information;

4. an analysis of the human resources available and needed to accomplish the above tasks;

5. an analysis of costs and required solutions to remove the constraints in accomplishing the above tasks;

6. recommendations on the most effective form of institutional and governmental organization to produce and deliver climate information for the public and private sectors.

VISION FOR THE FUTURE

The panel envisions an operational entity or entities that would create and deliver climate information products of benefit to society. This entity would not only concentrate resources on the needed modeling activities but would also establish and maintain a climate observing system to develop and test climate models and make available climate information for public and private use. Researchers would interact with this group to develop, diagnose, and improve models and observational systems. This interaction would provide the research community with otherwise unobtainable resources and result in enormous benefits and a sound foundation for future improvements.

1

Questioning the Effectiveness of
U.S. Climate Modeling

Over the last 30 years, data have revealed that the global climate is
naturally variable on many time scales (ranging from years to centuries
and longer) and may be changing in response to anthropogenic inputs of
radiatively active gases. The public and private sectors have become increasingly concerned with the potential impacts of this change. In response, the research community has been working to provide societally
beneficial information (e.g., NRC, 1999d), which has led to dramatic increases in requests for climate information products, particularly those
that can be used to understand the impacts of climate changes and evaluate strategies for dealing with them. Mathematical models based on the
laws of environmental physics and sound scientific measurements are the
primary tool that can be used to provide these products. When coupled
with descriptions of the social, political, technical, and economic impacts,
these models have the potential to predict future socio-economic changes
resulting from climate changes (e.g., Nordhaus and Boyer, 2000).

Recognizing the societal importance of climate modeling, the National Research Council published a report entitled *Capacity of U.S. Climate Modeling to Support Climate Assessment Activities* (NRC, 1998a). This
report evaluated the allocations of resources to high-end climate modeling and whether these resources were being used effectively. The full text
of the Executive Summary of this report is given in Appendix B. The
report concluded that, while small- and intermediate-scale climate modeling in the United States is effective and enjoying both national and
international prominence, climate modeling at the highest end is lagging
because of a lack of coordination among agencies, a lack of human and

computer resources devoted to the highest end of modeling, and a lack of an integrated national strategy for high-end modeling efforts. It further concluded that (1) the United States lags behind other nations in its ability to model long-term climate; (2) it is inappropriate for the United States to depend on other countries to provide high-end climate modeling capabilities; and (3) this dependence should be redressed by improving the capabilities within the United States. Finally, it concluded that "to facilitate future climate assessments, climate treaty negotiations, and our understanding and predictions of climate, it is appropriate to develop **now** a national climate modeling strategy that includes the provision of adequate computational and human resources and is integrated across agencies."

This present study focuses on the challenges posed in the 1998 report (NRC, 1998a) as specified in the statement of task given to the panel (Box 1-1).

To address these tasks, the Panel on Improving the Effectiveness of U.S. Climate Modeling met three times and held a one-day workshop

---

Box 1-1
Statement of Task

The purpose of this study is to provide relevant federal agencies and the scientific community with an assessment of the nation's technical modeling needs and a vision of how government, interacting with the rest of the scientific community, can optimize the use of modeling talents in the United States. This study will thus address the challenges posed in the Climate Research Committee's 1998 report, *Capacity of U.S. Climate Modeling to Support Climate Change Assessment Activities*. In pursuit of these objectives, the panel will produce a report that:

1. Examines the major types of climate modeling, paying particular attention to both the similarities (e.g., potential synergisms) and unique characteristics of each. Specific issues to be addressed include: model construction and testing, data input and archival, ensemble simulation, interrogation and diagnostics, evaluation, and operational utilization.

2. Describes the computational and human resources required to effectively conduct climate modeling in the United States to meet the needs of the climate applications, policy, and scientific communities. This evaluation will include consideration of shifts in computational architectures and potential for and cost of improvements in model codes. It will also consider the utilization of common climate modeling tools, protocols, and data, and the availability of cooperative opportunities between different scales of modeling effort and institutions.

3. Quantitatively assesses the computational and human resources that are presently directed toward climate modeling in the United States.

4. Describes ways in which the efficacy of the U.S. climate modeling enterprise might be improved, given the current needs and resources. The report will define a set of issues that are fundamental to the enhancement and sustenance of climate modeling in the United States.

with approximately 50 participants from the climate modeling community. This workshop was designed to gather information on community perceptions of the current state of climate modeling and possible responses (Appendix F). To quantitatively assess the computational and human resources presently directed towards U.S. climate modeling, two surveys were developed (Appendixes C and D). One survey was sent to large[1] and intermediate-size[2] modeling centers and one was sent to modelers who conducted workstation[3] scale modeling ("small" in the words of NRC, 1998a). Survey responses are tabulated in Appendix E.

Issues related to climate modeling have been the focus of a number of recent reports (summarized in Appendix H). The key issues and concerns arising from these reports that helped structure this study include:

1. the lack of adequate access to high-end computing by the climate modeling community;
2. the scarcity of human resources applied to computational and scientific research problems;
3. the difficulty of matching the financial rewards offered by private industry;
4. the lack of appropriate software available to optimize performance on the new generations of massively parallel computers;
5. the lack of software standards and protocols for building different climate models and the absence of uniform computer and observational data-archiving standards, which inhibits exchange of useful information;
6. the need for uniform criteria with which to judge climate models;
7. the need for widely available standard software tools to diagnose and compare climate model output;
8. the need for a strong interaction between observations of the climate system, research into fundamental climate processes, and integrative climate modeling.

Traditionally, climate modeling has been devoted to perfecting the understanding of the climate system. This has been done by the climate research community through competitive proposals that required little interagency coordination and no general strategy for success. Increasingly, however, modeling efforts have been directed towards the produc-

---

[1]In this document a large or high-end effort is one using a global, coupled T42 (2.8° × 2.8°) atmospheric / 2° × 2° oceanic model (or finer resolution) for centennial-scale simulations of transient climate change.

[2]In this document an intermediate center is one using a global, stand alone atmospheric climate model at T42 (2.8° × 2.8°) resolution.

[3]In this document a small modeling center is that which uses a global, stand alone atmospheric climate model at R15 (~4.5° × 7.5°) resolution.

tion of climate information products in response to the societal demands.[4] The interaction between researchers and users is then critical to determining what ought to be predicted and the limits and uncertainties of the predictions (Hooke and Pielke, 2000). The prediction products would then provide information and support decisions about agriculture, energy, health, transportation, food aid, disaster response, and other climatically influenced activities. This desire for information products and climate change assessments has been a primary motivator for increasing the accuracy of climate model outputs, particularly those modeling the climatic response to anthropogenically produced, radiatively active constituents. Similarly, advances in the ability to predict seasonal-to-interannual climate variations associated with the El Niño/Southern Oscillation have led to public and private demands for skillful predictions and for research on better ways of using the information (NRC, 1999c).

Increased societal demands for climate information products have had significant impacts on the research community, which, both by limited capacity and by culture, is increasingly unable to respond to these demands. Because the need for climate model products affects the current state of climate modeling science, a discussion of the impacts of these demands on the research community is included in this report. This situation is organizationally similar to that in the weather arena, where daily and weekly forecasts are provided for widespread public and private use by a service organization dedicated to that task, rather than by solicitation of proposals from the weather research community. The production of operational products in the weather community provides an organizational and institutional paradigm that can be applied to the climate situation.

This report will analyze the present capability for climate modeling in the United States, the current ability to respond to assessment, and prediction requirements. It will describe the new requirements being placed on high-end climate modeling and discuss the computer, human, and organizational resources needed to respond to these requirements. This is followed by findings and recommendations designed to improve the ability of the climate community to meet these new challenges. The report ends with a vision for how climate research, global observations, and comprehensive climate modeling could be combined for the benefit of science and society.

---

[4]The term "demand" is not used in this report in the strictest economic sense. Additionally, references to societal demands, imperatives, and needs in this report are based on panel members' individual experiences and interactions with specific societal uses as well as the perceptions of potential users of a broad range of climate modeling products and services. These terms also reflect the increased reliance of regional and national assessments on climate models. This latter point is addressed further in chapter 4.

2

Climate Models, Observations, and Computer Architectures

2.1 MODEL CONSTRUCTION

The internal cycling of the different elements in the Earth system and their interaction and feedback with the other elements ultimately creates climate. Creating models that characterize climate requires information about the various subcycles in order to characterize the interactions and feedbacks (Box 2-1) and the resulting amplification or dampening influences on the climate. Individual boxes and arrows (Figure 2-1) are useful for demonstrating how the integration of the individual elements of the environmental system ultimately produces a model for climate.

In general, each component of the climate system (atmosphere, ocean, land, ice) is modeled separately. Oceanographers build models inputting such information about the oceans as bottom and coastal topography, total amount of water and total salt content. In response to time-dependent inputs of freshwater (as rainfall and river runoff), momentum fluxes, and heat fluxes at the surface the models calculate the distribution of salinity, temperature, momentum (currents), density, and sea ice in the oceans over time.

Similarly, atmospheric scientists build models of the atmosphere incorporating surface geography and orography and the amount and distribution of gases in air (N_2, O_2, CO_2, H_2O and the more minor gases). In response to the input of radiation from the sun, to boundary conditions of specified time-dependent sea-surface-temperature (SST) and land surface

13

Box 2-1
Feedbacks in Climate Models

Feedbacks in the climate system occur when the output from one component is input into a second component, which then generates an output altering the first component. For example, increasing ambient air temperatures cause higher sea-surface temperatures, which result in decreased CO_2 dissolution into the oceans, leading to higher atmospheric CO_2 concentrations, which increases ambient air temperature. Climate models are constantly adjusted to account for the multiple non-linear feedbacks, which are common in nature.

---

type, and to the distribution of industrial and other anthropogenic aerosols and gases, the atmospheric models calculate:

1. the time-dependent distribution of temperature and pressure,
2. momentum (winds),

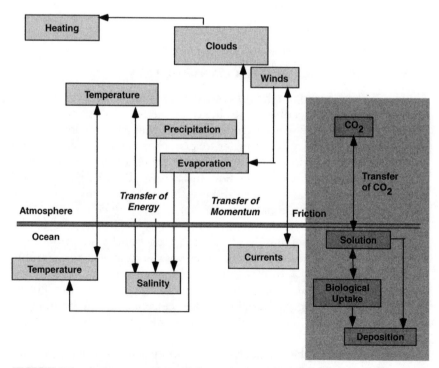

FIGURE 2-1   A representation of the major coupling mechanisms between the atmosphere and ocean subsystems. The processes in the shaded area are being developed offline. (Figure adapted from McGuffie and Henderson-Sellers, 1997. "A Climate Modelling Primer." Figure 1-3 John Wiley and Sons, New York.)

3. precipitation,
4. cloudiness,
5. humidity,
6. radiation within the atmosphere and at the lower boundary,
7. the total infrared radiation escaping to space.

Land specialists build models containing factors influencing the distribution and runoff of water on the surface, soil moisture, the growth of vegetation (which in part determines land albedo, the amount of evapotranspiration, and the uptake of carbon dioxide) and include the geography and topography of the land surface. In response to time-dependent inputs of radiation, water, gases, and winds from the atmosphere, the models calculate the time-dependent temperature, vegetation type, soil moisture, snow cover, and runoff. Modeled biogeochemical cycles describe the exchange of carbon, constituents of nitrogen and sulfur between the biosphere, ocean, and atmosphere.

These different component models are incorporated into the climate models, which use the interaction and feedbacks among the different cycles to describe the temporal state of the climate. For example, fluxes of water and momentum delivered from the atmosphere to the ocean in part determine the sea-surface temperature, which in turn influences the atmospheric temperature. The coupling of these components forms a global climate model.

One of the goals in climate modeling is to calculate the properties and evolution of the climate system in response to forcings, which represent the external changes in the components of the system, affecting climate. Changes in solar output, the effective changes in solar radiation reaching Earth caused by alterations in the orbital parameters of Earth, and change in volcanic emissions of aerosols are clearly forcings external to the climate system. External forcings include those that are specified even when they are internal to the climate system. Thus, specified changes in the chemical composition of the atmosphere leading to alterations to the incoming or outgoing radiation are forcings within the climate system, but because they are specified rather than calculated, they are treated as external forcings. Forcings have both a natural and anthropogenic component (Box 2-2).

2.2 OBSERVATIONS AND CLIMATE MODELS

Most of the information about the climate system derives from observations taken for purposes other than climate. For example, operational measurements of atmospheric temperature and humidity are routinely taken by a variety of means for the purpose of weather forecasting. Because the period of interest is very short (up to about 10 days), constant

---

Box 2-2
Humans and Climate Forcings

Over the past 30 years, there has been increasing interest related to the role of humans in climate. Humans produce large amounts of gases and aerosols such as CO_2, soot carbon, and SO_4^{2-} aerosols that are capable of altering the radiation budget both directly and indirectly. Additionally, land-use practices alter the distribution of vegetation on continents thereby changing the albedo, or reflectivity, of the land surface. The impact of human activities on climate is strongly debated in both the scientific and political communities. One of the goals of climate modeling is to quantify the role of humans in forcing climate change.

---

changes in the models and observing systems are made to improve the forecasts, but any change introduced into the system will cause a discontinuity that can be confused with a climate change. Also, cost saving measures for weather observations can put long time series, of major value for climate, in jeopardy. It was a conclusion of NRC (1999b) that our current ability to adequately document climate change was compromised by an observing system unsuitable to the task.

The basic properties required of a climate monitoring system as enumerated in NRC (1999b) were:

1. Changes to an observing network should be assessed in terms of the effects on climatic time series.

2. Any replacement instruments should be overlapped with the old ones for an appropriate period of time.

3. Metadata which documents the instruments and procedures should be kept and archived along with the data.

4. Data quality and homogeneity should be assessed as part of routine operating procedures.

5. The data should be used in environmental assessments of various types so that it will be constantly examined.

6. Historically important time series within the observing system should be maintained and protected.

7. Data poor or otherwise unknown or sensitive regions should receive special priority.

8. The entire system should be designed with climate and weather requirements in mind.

9. Commitment to old systems and a transition plan from research to operations needs to be a part of the system.

10. Every effort should be made to facilitate access and use of the data by national and international users.

A climate observing system is inadequate when it either fails to measure climatically important quantities or when the measured quantities do not satisfy the 10 properties above. By this standard there is not an adequate climate observing system. The existing weather observing system does not satisfy the 10 principles. Some important climatic quantities, such as subsurface ocean temperature and salinity, land soil moisture, and the concentrations of specific atmospheric species such as the hydroxyl radical, are either measured inadequately or are not measured at all.

An effective integrated system for producing and delivering climate information has as one of its major elements the climate observing system. To the extent our vision of coupled climate observations, high-end modeling, and research is the proper one, and to the extent that the relationships between the elements are as important as the elements themselves, one cannot solve the high-end modeling problem without solving the sustained climate observations problem. Because observations cost an order of magnitude more and the infrastructure to maintain sustained observations is again an order of magnitude more than any likely modeling infrastructure, the problem of producing and delivering climate information is, to first order, one of creating and maintaining a climate observing system.

Improving Climate Models with Observations

Climate models are built using our best scientific knowledge of the processes that operate in the atmosphere, ocean, land, and cryospheric systems, which in turn are based on our observations of these systems. Climate models and their gradual improvement therefore arise from the totality of the research enterprise, which can be diagrammed as shown in Figure 2-2. The climate system is observed, and on the basis of these observations, physical processes (e.g., the radiative and other thermal processes that determine the temperature) and the large-scale structure of the component systems are understood. Models of the component systems are constructed and compared to observations. When disagreements between models and observations are noted, model processes are improved, perhaps by performing field studies devoted to a single process (e.g., clouds) or perhaps to acquire a detailed set of observations of a set of interacting processes by which to improve the details of component models. The state and accuracy of climate models depends on the state of all the elements of Figure 2-2. It is in this sense that climate models contain our accumulated wisdom about the underlying scientific processes and can be no better than our observations of the system and our understanding of the processes that interact to form the climate.

Weather forecasting provides an example of a process for improving climate modeling based on the interactions of the elements in Figure 2-2.

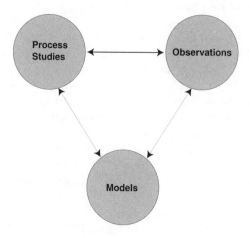

FIGURE 2-2   Modeling paradigm.

Sustained global observations of the atmosphere have been taken for the last 50 years in support of weather forecasting. These observations are assimilated into a global atmospheric model, and an analysis of the current state of the atmosphere is performed in order to initialize a weather forecast. The skill of the forecast is primarily determined by the accuracy of this initial analysis, which in turn depends on the coverage of the observations, the quality of the observations, the accuracy of the models, and the accuracy of the data assimilation scheme.

The sequence of twice (or four times) daily analyses is normally what researchers consider "data," but it is as close to an optimal blend of observations and model output as current science allows. Thus, weather prediction is the arena in which observations used to define climate are taken and archived, ultimately forming the basis for our knowledge of the climate of the interior of the atmosphere. Hollingsworth et al. (1999) have articulated the advantage of linking observations with forecasting; "The forecast centre's ability to compare every single instantaneous observational measurement with a forecast of that measurement is a powerful scientific resource." This resource can be employed to improve the parameterization of processes in models, and to gauge the adequacy of the observing system ultimately improving the forecast models, and the skill of weather forecasts. A similar method to improve the parameterizations of processes consists of confronting models with data taken from field programs designed to illuminate physical processes not adequately resolved by routine weather observations. Combinations of the two methods can be used.

2.3 PURPOSES OF CLIMATE MODELING

Climate models are employed for a number of different purposes. Climate models as close to comprehensive as possible can be used to simulate the climate system. The model can be run for a long time and its annual cycle and variability about that annual cycle assessed. Once the model has proven itself in this simulation mode, simulations can be performed on the climate model, which clearly could not be performed on the climate. For example, the climate model can be used to simulate the response of the climate model to external changes, such as to the solar constant or to volcanic eruptions. The most common type of assessment is to increase the concentrations of carbon dioxide in a model according to the measured increase over the last 45 years and inferred increases before that. The response of the climate system to these changes, with or without external changes, can then be compared to the observed changes in order to infer the causes of whatever climate change has occurred. The climate model can then be run into the future to either examine the response to projected increases in the concentrations of radiatively active constituents or, in the case of comprehensive climate models, to examine the response to projected emissions of radiatively active constituents.

Coupled climate models can be used to probe for predictability on short time scales (months to years) when some data exists. Ideally, a coupled assimilation is performed using past data to achieve an initial analysis of the state of the coupled atmosphere-ocean-land system, and the system is allowed to run freely to give a (retrospective) prediction of the state of the climate. Because the data exists for the prediction period, these hindcasts can be used to determine what the skill of prediction would have been over a long series of forecasts if they had taken place prospectively instead of retrospectively. Once skill in hindcast mode is demonstrated, coupled climate models can be used to predict the climate a season to a year in advance.

Coupled climate models can be used to probe for predictability in the climate system on longer time scales when no observational data exists. A long simulation can be run and the model output treated as if it were true observational data. The method described above can then be used on the simulated model output, perhaps sampled at points for which observations could exist, and simulated predictions could be made. Comparing the predicted state of the climate to the simulated "true" future state (again perhaps sampled at the points of a presumed observational system) could give a rough guide to the existence of predictability.

A similar method could be used to guide the design of proposed observing systems. The state of the model climate is completely known in both space and time. The ability of various proposed observing systems to accurately describe the model climate could be tested by sampling the model system at the proposed observation points with errors characteristic of actual observing systems.

Because the details of the climate are affected by weather noise that needs to be averaged out in climate assessments, climate ensembles must often be run. Ensembles usually include 10–30 members (see discussion in Section 5.1).

An important recent use for models is to downscale information from the resolution at which the climate is simulated by global models, to a much smaller region at much higher resolution in order to capture the local characteristics of the specific region. Output from the global climate model is used as lateral boundary conditions for much higher resolution regional atmospheric models that then capture the local peculiarities of terrain and orography and, ideally, return details on local weather and weather changes under different climatic regimes. Questions have arisen as to the consistency of this method because it does not allow the back reaction from the small to the large-scale.

2.4 COMPUTER ARCHITECTURES IN SUPPORT OF CLIMATE MODELING

Climate and weather modeling require enormous computing capacity and capability. Over the next few years the two supercomputer architectures that will provide this computing power are vector parallel processor machines (currently manufactured primarily in Japan) and microprocessor-based massively parallel computers (currently manufactured primarily in the United States) (Box 2-4).

In a few years, high-end computer systems expected to be available internationally can be divided roughly into the following categories:

1. Clusters of nodes, each node having multiple shared-memory processors (SMP) running some variant of the Unix operating system. These server systems are composed of commodity processor and memory chips on nodes interconnected by custom-designed networks.

2. Loosely integrated clusters of PCs running the Linux operating system. "Linux clusters" are based on commodity PCs using standard PC processor and memory chips and interconnection networks built of commodity parts. These build-your-own systems are considerably cheaper per peak gigaflop than those in category 1, but they tend to be software poor and often suffer from reliability problems.

3. Computers based an innovative and promising new architecture are being produced in limited quantities by TERA, a small U.S. company. The TERA computer uses specially designed processors that use dataflow concepts to support fine-grain parallelism; each processor supports up to 128 threads of execution to hide the latency of outstanding

---

Box 2-3
A quick tour of the computer terminology used in this report

The heart of any computer is its processing element (PE; also known as 'processor' or 'central processing unit'). The PE can have multiple *pipelines*, each of which can evaluate *floating-point* operations in a sequence of stages, one stage per clock cycle. Each PE may have its own memory unit that only it can address directly, as in a normal personal computer (PC), or it may share access to a memory unit with several other PEs (see *node*). The PE may read from and write to memory directly, as in traditional *vector* processors like the Cray C90 and NEC SX-4, but this requires very high memory bandwidth. *Cache*-based *microprocessors* use memory chips that are very slow compared to those used in vector machines. To compensate, microprocessor-based computers place one or more levels of *cache* between the processor and memory units.

Cache is a special memory unit that provides very fast access to and from the PE. If the datum needed by the PE resides in the highest-level (L1) cache, it can be retrieved from cache to the PE in only a small number of clock cycles. If not, a "cache miss" occurs and the request must then go to the next level (L2, etc.) of cache. If the datum is not in any level of cache, the request goes to the memory unit on the node. Each successive level of cache is progressively larger but takes longer for the requested datum to reach the PE.

A node is a collection of two or more processors grouped together so that they share access to a common memory unit. This set of PEs is said to have uniform memory access to the entire unit of shared memory. A node has a single-system image because only a single copy of the *operating system* needs to be stored in memory to be shared by all the PEs.

A collection of nodes connected by a network is said to have distributed *shared memory* if the computer's architecture supports direct reading or writing by a PE of data stored in memory units on the other nodes. Such a computer also has a single-system image. If inter-node communication can only take place via messages that must be handled by the PEs, then the computer simply has distributed memory. In either case, because the time to access off-node memory is much longer than on-node memory and even varies with the "distance" between the nodes in the network, off-node memory access is termed non-uniform memory access .

---

memory references. The San Diego Supercomputer Center owns an 8-processor TERA that has been shown to perform extremely well on problems involving gather-scatter memory operations (Oliker and Biswas, 1999).

4. Tightly integrated VPP systems with distributed shared memory using high-performance processors, memory, and interconnection network custom-designed to work together. Manufactured by Japanese vendors, these systems are widely used in countries other than the United States, where political pressures prevent their sale. VPP systems may become available again from once preeminent Cray Research Inc. (CRI).

Box 2-4
Supercomputing Architectures

The largest collection of PEs supporting a single-system image, whether it is a node or a collection of nodes interconnected by a network, is termed a symmetric multi-processor or shared-memory processor (SMP) system. An SMP cluster is a collection of SMP systems that are connected by yet another network.

It has become customary to use the terms "Vector Parallel Processor" (VPP) and "Massively Parallel Processor" (MPP) to refer to computers based on custom-designed vector processors and commodity microprocessors, respectively. This can be confusing because both classes of computers actually use multiple processors operating in parallel to compute different parts of the problem. Furthermore, both types of computers employ networks that interconnect the processors to permit the exchange of information. The distinctions between VPPs and MPPs are manifested by (a) the number and power of the processors and (b) the manner and speed with which data can be moved from memory to the processors. Vector processors are more than an order of magnitude faster than the fastest microprocessors, so while VPPs might have 16-32 PEs, an MPP with comparable peak speed might have 512-1024 PEs. In addition, the ratio of sustained-to-peak performance for a typical parallelized code is in the range of 30–40% for VPPs but only ~10% for MPPs. It is the large number of processors that gave rise to the term "massively". The power of vector processors depends critically on having a very fast network connecting the PEs to memory, capable of delivering one (or more) operand(s) per clock cycle. The commodity memory chips used with microprocessors are, by comparison, quite slow. Cache must then be used to keep data "closer" to the PEs. To obtain good performance requires that the programmer explicitly design the code to divide the problem up into very small "pieces", each of which will fit in L1 cache. Because present-day SMPs are built of cache-based microprocessors, "SMP cluster" supercomputers are often referred to as MPPs, and may be referred to as such in this report.

After falling on hard times due to the shrinking marketplace for supercomputers, CRI was purchased by SGI, and later sold to TERA, which adopted the name "Cray Research" as its own. Prior to being acquired by TERA, CRI was designing the SV-2, a scalable vector successor to the T3E. Time will tell whether TERA/CRI management will decide to produce the SV-2.

What are the implications for climate, atmosphere, and ocean modeling of such limited options? Let's look more closely at the characteristics of clusters of processors like those in categories 1 and 2 above.

1. Commodity processors are very slow (a factor of 10 or more) compared to custom-designed vector processors. This means, of course, that proportionately more commodity processors must be used to get the same

peak performance. Getting the same *sustained* performance may or may not be possible, depending on the scaling of the model as more processors are employed.

2. A much larger network is needed to connect the processors. To hold down the cost of the network, commodity components are used, although the performance of the network suffers. The success of the CRI T3E (an SMP type of computer) has been largely because of its very fast, custom-designed interconnection network.

3. Finally, the large size and reduced performance of the interconnection network compounds the slowness of the memory chips, resulting in long delays (high latency) in obtaining data from outside a node. Additional levels of cache introduced in an attempt to hide the latency further complicate the job of the programmer and software engineer. The intrinsic ability of the TERA computer to hide latency without using cache is intriguing but needs to be tested on a variety of applications.

These shortcomings of commodity-based supercomputers are particularly detrimental in the context of climate, ocean, and atmosphere modeling. An important characteristic of the oceans, and therefore, of the climate system is the slow rate of change of the deep ocean in response to changes in surface forcing. In modeling, this leads to very long integrations, measured in both simulated and actual. There are several reasons for this:

1. In order that "climate change" due to changes in forcing be distinguishable from "model drift" caused by starting the ocean model from initial conditions that do not represent a quasi-equilibrium state of the ocean *model*, the model must be integrated for thousands of simulated years until an equilibrium state is reached. Acceleration techniques (Bryan, 1984) have been developed to speed the approach to equilibrium in the deep ocean, but even with acceleration the model may have to be run for hundreds of (unaccelerated) "surface years."

2. Once a suitable initial condition for the ocean model is obtained, the model must be run for centuries in order to sample the broad spectrum of time-scales present in the natural variability of the ocean.

3. In scenarios for global climate change due to anthropogenic influences, such as global warming, effects in the models become perceptible over decades to centuries, depending on the assumed rate of change in the forcing.

Completing such long (in simulated time) runs in an acceptable amount of wall-clock time places limits on the spatial resolution that can be used. The longer and more numerous the runs, the coarser the resolution must be. One might think that the wall-clock time could be reduced arbitrarily by simply applying more processors. But when the number of grid points per processor, known as the "subgrid ratio," becomes too small, various inefficiencies come into play, reducing or eliminating entirely the benefit of adding more processors.

The principal inefficiency arises from the incomplete parallelization of the code. It is quantified by Amdahl's Law (Hennessy and Patterson, 1990), which can be stated as follows. Let $T_{tot}(1)$ be the total wall-clock time required to complete some calculation on a single processor [processor element (PE)]. Let T_s and T_p be the times to run the serial and parallel portions of the code, respectively, on one PE. Then $T_{tot}(1) = T_s + T_p$. Now on a system with N PEs, the time to run the parallel portion of the code is reduced to T_p/N, while the serial part still runs on one PE in T_s. Thus, $T_{tot}(N) = T_s + T_p/N$ and the speed-up on N processors is $S(N, f_s) = T_{tot}(1)/T_{tot}(N, f_s) = 1/(f_s + f_p/N)$. Here $f_s = 1 - f_p = T_s / (T_s + T_p)$ is the fraction of single-PE time spent on serial code. For large values of N, $S(N, f_s) \to 1/f_s$, independent of N. A well-parallelized code might have $f_s \sim 1\%$ and a highly parallelized code might have $f_s \sim 0.1\%$, giving asymptotic speed ups of 100 and 1000, respectively. This speedup limit is a property of the code, not the computer, and cannot be exceeded regardless of the number of PEs applied (Plate 1).

The implications of Amdahl's law are serious when one considers the difference in performance between custom-designed vector and commodity cache-based PEs. Not only do vector parallel processor (VPP) systems have much higher peak performance per PE (~ 3 Gflops/PE for the NEC SX-4) than do cache-based distributed memory machines (~ 0.5 Gflops/PE on the SGI Origin 2000) but the sustained performance is also typically a much higher percentage of peak performance on VPPs (~ 30–40%) than on SMPs ($\sim 10\%$). Thus, the *sustained* per-processor floating-point performance ratio is roughly a factor of 20 or more in favor of the vector processors.

An example may help illustrate the consequences of Amdahl's Law. Assume that some simulation, such as a century-long run with a medium-resolution ocean model, requires 10^{15} floating-point operations to complete and you want it finished overnight (12 hours). This implies an aggregate sustained rate of 23 Gflops (Table 2-1).

Obtaining the degree of parallelism corresponding to $f_s = 0.01$ in atmospheric and oceanic codes is challenging. The modest speedup needed by the VPP system is attained easily for $f_s = 0.01$ and trivially for $f_s = 0.001$. The much larger speedup of 463 required for the SMP system cannot be attained with any number of processors applied to a code with $f_s = 0.01$ because $S(N, f_s) < 1/f_s = 100$. The SMP system is hard pressed even with a much more highly parallelized code ($f_s = 0.001$): To attain a speedup of 463 requires 850 processors, an efficiency of 54%. The network needed to support 850 PEs is much larger; to control its cost it must be built from

Box 2-5
Measuring the Sensitivity of Peak Versus Sustained
Performance for Specific Applications

Fast processing speeds quoted for highly parallel machines, such as those listed on the website of the top 500 supercomputers (http://www.top500.org/), are often determined using a collection of matrix routines making up the benchmark software Linpack. The difficulty in using Linpack to compare computing performance for climate modeling applications is that each Linpack routine represents a small computational kernel that can be optimized for MPP systems, whereas a climate modeling code must represent a diverse set of physical processes and therefore lacks a kernel with a comparable degree of parallelism. To enable a more realistic benchmark, the National Aerodynamic Simulation (NAS) Parallel Benchmark Suite (Bailey et al., 1991) was developed (Figure 2-3).

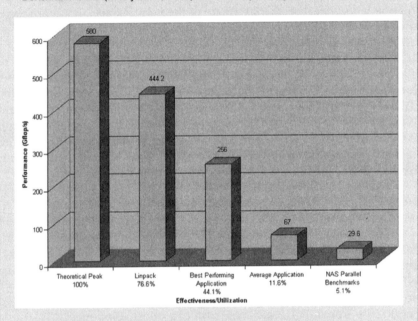

Figure 2-3 Measured performance of applications on the 644-processor Cray T3E at the National Energy Research Scientific-Computing Center (NERSC). This figure shows the sensitivity of performance to specific applications. These performance curves show that the Linpack benchmark does not represent a general application environment accurately. The average application at NERSC performs at 11.6 % of the theoretical peak. In U.S. climate-modeling centers a 10% performance goal is often set. Thus, to achieve *sustained* speeds of 10–100 Tflops, *peak* speeds of 100–1000 Tflops would be required. Sustained-to-peak ratios in excess of 33% are common on the Japanese VPP computers, which coupled with their much higher single processor speed leads to a performance-usability gap between MPP and VPP systems.

TABLE 2-1   Serial and Parallel Performance of Models on
Microprocessor and Vector Processor Systems

Computer type	VPP	SMP
Sustained rate [Gflops/PE]	~ 1	~ 0.05
$T_{tot}(1)$ [hours]	278	5556
Speed-up factor required = $S(N, f_s)$	23	463
N required if f_s = 0.01	30	no solution
Efficiency = $S(N, f_s=0.01)/N$	77%	—
N required if f_s = 0.001	24	850
Efficiency = $S(N, f_s=0.001)/N$	96%	54%

cheaper, slower parts than a network to support 24–32 vector processors. As pointed out earlier, the large size of the network and the slowness of the network components conspire to reduce further the effectiveness of SMP clusters compared to VPPs.

Strictly speaking, Amdahl's Law can be applied rigorously only to systems in which a single processor has uniform access to all of the memory required to hold the same problem that is run on N processors. This is the case in traditional shared-memory VPPs like the Cray C90 and NEC SX-4, but it is not so in distributed-memory SMPs like the SGI Origin 2000 or the IBM SP. The latter have non-uniform (much slower) access to memory that resides on other nodes, compared to the access rate to memory on the processor's node, resulting in further slow down.

Another major inefficiency that affects codes on parallel computers is load imbalance: different processors have differing amounts of work to do, so that some sit idle while others are overburdened. This is a problem for atmospheric General Circulation Models (GCM) that arises from differing amounts of work over land and ocean in the day and night hemispheres and at different latitudes. In ocean models land points must be eliminated from the computational domain to the greatest extent possible so that processors do not sit idle. If the component models of a coupled model run in parallel, load imbalance can cause components to sit idle while waiting to receive information from another component.

**Box 2-6
The Load Balancing Issue**

To achieve good performance with massively parallel computers each processor should be given an equal "load" of the total computational work. If there is a load imbalance where one or more processors compute significantly more of the load, then a significant fraction of the processors may lay idle during a given computation, reducing both efficiency and overall performance.

These facts regarding parallel scalability are reflected in the performance achieved for important, widely used climate and weather codes. Table 2-2 summarizes the performance of several such codes, including

TABLE 2-2 Comparison of Capacity and Capability Between Vector Processors and Massively Parallel Processors in Supercomputing

Model		Single Processor (serial execution)	Multiple Processor (parallel execution)	
Processor type	Micro	Vector	Micro	Vector
IFS (ECMWF) [1]	[1]CRI [2]T3E [3]600 Mhz [4]1 PE [5]2.14 fdpd [6]N/A	[1]Fujitsu [2]VPP 5000 [3]9.6 Gflops peak [4]1 PE [5]102.5 fdpd [6]N/A [7]48	[1]CRI [2]T3E [3]600 Mhz [4]1408 PEs [5]1800 fdpd [6]N/A	[1]Fujitsu [2]VPP 5000 [3]9.6 Gflops peak [4]98 PEs [5]7100 fdpd [6]N/A [7]56
MC2 (Canada) [2]	[1]SGI [2]Origin 2000 [3]250 Mhz, 4 MB [4]1 PE [5]80 Mflops [6]16%	[1]NEC [2]SX-5 [3]8 Gflops peak [4]1 PE [5]3500 Mflops [6]44% [7]44	[1]SGI [2]Origin 2000 [3]250 Mhz, 4 MB [4]12 PEs [5]840 Mflops [6]14%	[1]NEC [2]SX-5 [3]8 Gflops peak [4]28 PEs [5]95200 Mflops [6]42% [7]48
GME/LM (Germany) [3]	[1]CRI [2]T3E [3]600 Mhz [4]1 PE [5]60 Mflops [6]5%	[1]Fujitsu [2]VP 5000 [3]9.6 GF/s peak [4]1 PE [5]3000 Mflops [6]31% [7]50	[1]CRI [2]T3E [3]600 Mhz [4]1200 PEs [5]? [6]?	[1]Fujitsu [2]VP 5000 [3]9.6 Gflops peak [4]? [5]? [6]? [7]50
MM5 (US) [4]	[1]Compaq [2]Alpha Server [3]667 Mhz [4]1 PE [5]360 Mflops [6]27%	[1]Fufitsu [2]VP 5000 [3]9.6 Gflops peak [4]1 PE [5]2156 Mflops [6]22% [7]6	[1]Compaq [2]Alpha Server [3]667 Mhz [4]512 PEs [5]45317 Mflops [6]6.6%	[1]Fujitsu [2]VP 5000 [3]9.6 Gflops peak [4]20 PE [5]27884 Mflops [6]15% [7]16

The entries in the table are: [1]manufacturer; [2]model; [3]processor characteristics; [4]number of processing elements (PEs); [5]sustained rate obtained with model (units are either "fdpd" = forecast days per day, or "Mflops"), [6]sustained rate as % of peak performance rate; and [7]ratio of sustained rates *per processor* of vector processor(s) to microprocessor(s).

[1] Personal Communication, David Dent, ECMWF, Great Britain.
[2] Personal Communication, Steve Thomas, NCAR, USA. Also see: M. Desgagne, S. Thomas, M. Valin, "The Performance of MC2 and the ECMWF IFS Forecast Model on the Fujitsu VPP700 and NEC SX-4M", to appear in the *Journal of Scientific Programming*.
[3] Personal Communication, Ulrich Schaettler, Deutscher Wetterdienst, Germany. Also see the paper: D. Majewski, *et al.*, The Global Icosahedral-Hexagonal Grid Point Model (GME) Operational Version and High-Resolution Tests, ECMWF Workshop on Global Modeling, 2000.
[4] Personal Communication, John Michalakes, NCAR, USA. Also see the web site: *http://www.mmm.ucar.edu/mm5/mpp/helpdesk* and the paper: J. Michalakes, "The Same Source Parallel MM5" to appear in the *Journal of Scientific Programming*.

IFS, MC2, MM5, and LM/GME (A more detailed description of each code is provided in Appendix I). Each code has a best microprocessor and vector performance achieved for both serial and parallel execution. We then compare them using the ratio of the sustained performance *per processor* between vector and microprocessor machines for both serial and parallel execution. The implications of the different performance characteristics will be explored in the following sections.

3

State of U.S. Climate Modeling

An important task of this study was to quantitatively assess the computational and human resources presently directed towards climate modeling in the United States. To accomplish this goal two surveys were developed (Appendixes C and D). One of these surveys was sent to large and intermediate-size modeling centers and one was sent to small centers. After these surveys were drafted, a specialist in social surveying edited them to ensure that the information collected was as free of bias as possible. These surveys were sent to 50 modeling institutions and groups and 42 responses were received. The panel does not claim to have surveyed all groups or institutions operating small-scale modeling efforts. Because of the varied and extensive use of modeling in many areas of earth science, it would be extremely difficult to identify all of these small centers, thus, those responses that were received were taken to be indicative of smaller efforts. A good estimate of resources could be obtained for the largest centers because they were easier to identify, and all responded. Survey responses are discussed below and tabulated in Appendix E.

3.1 MODELS

The information collected on current modeling activities shows the robust and varied nature of climate and weather modeling in the United States. Smaller modeling centers enjoy a level of resources equivalent to what would have been considered supercomputer resources only a decade ago allowing them to run either in-house (regional and global) component models, component models from the larger centers, or a combination of the two. Smaller centers can even run coupled climate models but at coarse resolution (e.g., 800 km) or a higher resolution (300 km) for shorter time periods. Responses to the question about improvements that are being planned for models at all centers were varied, but most involved a mixture of increased model physics, dynamics, numerics, efficiency, and applicability. Many respondents also noted the desire to better incorporate new types of satellite and radar data.

In general, most respondents stated that their code was portable on platforms other than those on which they normally operated although some models required a moderate amount of optimization to ensure that they ran with minimal performance loss. Most centers release their modeling results to the wider scientific community without restrictions. A few centers freely release their data but stipulate that the results be used only for research purposes; others limit the release of modeling results to collaborators.

Large and intermediate-size modeling centers were asked whether there were plans to convert model code to run on massively parallel (MPP) architectures. Most institutions responded that this conversion had already taken place although those that have converted or are in the process of converting noted the difficulty in transferring certain models to an MPP architecture. Many respondents also noted that this conversion required significant programmer time and drained resources that could have been devoted to other activities. When asked for comment on the relative merits and hindrances of MPP versus VPP architectures, the majority of respondents preferred VPP architecture for the following reasons:

1. MPP systems are generally more difficult to program and require increased computer expertise. There are therefore significant training issues involved in the use of these systems. These difficulties are particularly significant for university centers as they often rely on graduate student labor that is characterized by high turnover.
2. Data assimilation and processing are more difficult on MPP systems.
3. VPP systems are more stable and reliable.
4. There are significant scalability problems on MPP systems.
5. There is a lack of compilers on current MPP systems that make these systems difficult to use.

Despite the difficulties with MPP systems some respondents felt that these systems had significant benefits over VPP systems (e.g., lower memory cost and increased aggregate CPU power).

3.2 COMPUTING[1]

Most small and many intermediate-size modeling centers either rely on the use of workstations or clusters of workstations for their modeling efforts or they collaborate with the larger centers and use their computational facilities. Larger-size modeling centers primarily rely on super-computers for their climate and weather simulations. Of the large modeling centers surveyed, half share their computational time with the wider community. The computing capacity of large and intermediate-size modeling centers are described in Table 3-1. This table also includes planned upgrades to existing systems.

When asked what upgrades would be incorporated if funds were available, the responses were varied (Table 3-1; Appendixes C and D), although the majority of centers noted the need for increased capabilities such as additional processors, nodes, and disk space or some combination. Some centers also noted the need for additional network bandwidth to more rapidly acquire data sets from remote sources. Some of the smaller centers, when asked what additional upgrades would be incorporated if funds were available, said they would prefer to devote any new funds to the purchase or enlargement of an existing PC cluster rather than pooling these funds to upgrade shared supercomputing resources.

Most centers (large, intermediate, and small) responded that computing capabilities were limiting the resolution and number of model runs and the production of real-time forecasting products. Although it is arguable that the desire will always be to produce a greater number of higher resolution, higher complexity model runs regardless of the available computational capacity, it is apparent that the ability to accurately model weather and climate at finer spatial and temporal scales is dependent on the ability to obtain a robust estimate of climate model uncertainty. This requires the analysis of a large number of cases and ensemble members per case. Increased model quality will lead to increased predictive skill and higher quality operational products for climate and weather prediction. Thus, the computational limitations noted in the survey are not only affecting current research activities and model development but also the production of outputs required for operational use.

It is important to note that, in addition to the need for additional computing capabilities, many respondents discussed the critical need for qualified scientists, modelers, and hardware and software engineers. This need is discussed more fully in the next section.

_____

[1] The information in Table 3-1 was accurate at the time that the survey results were assembled. Since then, information detailing the upgraded computing capabilities at NCEP was provided. The recently upgraded machine uses IBM's Power 3 Winterhawk-II technology, operating at 375mhz. The system has 2208 processors in 40 frames and has 512 compute nodes, with 2 GB memory per node.

TABLE 3-1   Computing Resources Located At Large Modeling Centers[a]

Institution[b]	Computer System	Processors	Last Upgrade
CIT-JPL	[1]Cray T3D/T3E [2]SGI Origin 2000	[1]512 [2]128	1999
COLA	[1]SGI Origin 2000 [2]Compaq ES40 [3]Compaq DS20	[1]16 CPUs [2]4 CPUs	1999
CSU	[1]SGI Origin 2000 [2]Octane	[1]10 [2]12	20% of inventory upgraded/year
FSU	[1]IBM SP2 with 9 nodes running on a fast interconnect bus 6 RS6000 model 260/270 series.	2 of the 260 series are dual processors, the remaining 4 units are 4 processor machines.	No major upgrades.
UCLA	[1]Compaq XP1000-cluster	[1]5	[1]1999
UH	[1]Cray SV-1 [2]SGI Origin 2000 [3]SGI Origin 2000 [4]SGI Origin 2000	[1]24/300 Mhz [2]32/250 Mhz [3]16/195 Mhz + 8/30 Mhz [4]4/180 Mhz	[1]March1999 [2]March 1999 [3]March 2000 [4]December 1999
UI	[1]NekoTech Jaguar 333Mhz [2]DCG Computers Viper 500 MHz [3]DCG Computers LX 533 MHz [4]DCG Computers LX 533 MHz [5]MicroWay Alpha 600 MHz	[1]1 [2]1 [3]1 [4]1 [5]1	[1]1995 [2]1997 [3]1997 [4]1998 [5]1999
IRI	[1]Cray J-9 [2]SGI O2000 [3]NEC SX-4B	18 and 16 [2]64 [3]2	4 years for Crays; nearly 1 year for Origin upgrade, just over 1 year for SX4.

Sustained System Performance	Central Memory / Secondary Disc Storage	Future Upgrades Planned
[1]10~50 Gflops	No information provided.	No information provided.
[1]2.5 Gflops [2]1.25 Gflops	[1]4 GB [2]4 GB/node Disk capacity: 2.3 TB (shared via gigabit-switch LAN)	None.
No information provided.	No information provided.	8-processor Origin in 2000 (Chance).
Unknown.	Each machine has approximately 2Gb of memory; 270's have ~ 50 GB of disk space Other machines have ~ 9Gb of disk space per machine.	
[1]2 Gflops	[1]2 GB/0.1 TB	None planned.
[1]28.8 Gflops [2]16 Gflops [3]6.2 Gflops [4]1.4 Gflops	[1]16.0 GB RAM/156GB [2]14 GB RAM/180 GB [3]4.5 GB RAM/36 GB [4]1.0 GB RAM/1 TB RAID5 (capacity is extended by Veritas HSM using a tape library with 13.6 TB capacity)	No information provided.
No information provided.	[1]64 M/9 G [2]128 M/18G [3]128 M/18 G [4]128 M/18 G [5]256 M/18 G	Three AlphaStation-type workstations in the next five years.
[1]1.5 Gflops [2]5 Gflops [3]2.5 Gflops	[1]32 Gbytes, 1.4Tb [2]16 Gbytes, 0.1Tb [3]8 Gbytes; 0.2Tb Additional mass store available (10Tb at LDEO, larger system at SDSC)	Crays will be replaced within the next year. New system not known yet.

TABLE 3-1   Continued

Institution[b]	Computer System	Processors	Last Upgrade
LANL	[1]SGI Origin 2000	[1]1024	[1]1999
NASA-DAO	[1]SGI Origin-2000 clusters	[1]Six 64-CPU machines, one 32-CPU machine	[1]2000
NASA-GISS	[1]SGI Origin 2000	[1]96	[1]1998
NASA-GSFC	[1]CRAY T3E/600 [2]DEC alpha 4100	[1]1024 [2]12	[1]2000 [2]1999
NCAR- M. Blackmon	[1]Cray C-90 [2]Cray J-90 [3]SGI Origin [4]IBM SP	[1]16 [2]16-20 [3]32, 64 or 128 [4]Variety of configurations	[1]Decomissioned in late 1999 [4]Spring 2000
NCAR- W. Washington	[1]CRAY T3E900 [2]SGI Origin [3]Origin 2000/128 [4]HP SPP2000 [5]IBM SP2 [6]Sun Starfire [7]DEC/Compaq [8]Alpha Cluster [9]Linux Cluster	Unknown.	Unknown.
NOAA-CDC	[1]Compaq AlphaServer DS10 [2]Sun Enterprise 4500 [3]Sun Ultra 60 [4]Sun Enterprise 450	[1]12 machines each with a single 466mhz Alpha 21264 processor [2]2 machines one with 8 UltraSparc II 400mhz processors, the other with 4 [3]6 machines each with 2 360mhz UltraSparc II processors [4]4machines each with 4 300 MHz UltraSparc II processors	[1]May 2000.

Sustained System Performance	Central Memory / Secondary Disc Storage	Future Upgrades Planned
[1]100 Gflops (theoretical sustained)[c]	[1]256 MB/processor or 256 GB/system	Unknown.
[1]~ 3-4 GFLOPS on each of the 64-CPU clusters	[1]16 GB central memory; disk space varies	Only minor upgrades planned.
[1]For mostly single-processor and ensembles of runs it is ~ 75 Gflops	[1]Central memory 20 GB/1000 GB	Upgrade to 128 processors and an upgrade of chip speed to the current state of the art as well as increased disk storage.
[1]40 Gflops [2]1 Gflop	[1]128 GB (mem) 750 GB (disk) [2]3.5 GB (mem) 1800 GB (disk), 20 TB mass strorage system	1and 2. doubling of capability for the current system in 2001 and another in 2003.
[1]~5 Gflops [3]~5 Gflops Both using 64 processors	Unknown.	New system procurement to be installed in early 2001
Unknown.	No information provided.	NCAR will soon be involved in procurement for a new system to be installed in early 2001.
[1]6.3 Gflops [2]3.6 Gflops [3]3.25 Gflops [4]3.6 Gflops LINPACK Gflops for aggregate of each system type.	[1]Each node has 512 MB/50 GB [2]4 GB on the 8-processor machine, 2 GB on the 4-processor machine [3]1 GB RAM on 3 machines, 2GB on the others [4]2 GB 2928 GB of disk storage shared by the Sun systems.	AlphaServer cluster will be upgraded, as faster processors become available (resources permitting).

TABLE 3-1 Continued

Institution[b]	Computer System	Processors	Last Upgrade
NOAA-GFDL	[1]SGI/Cray T932 [2]SGI/Cray T94 [3]SGI/Cray T3E (water-cooled chassis)	[1]122 [2]24 [3]128 450-MHz	[1]Upgraded to 26 processors in 1996; was de-rated to 22 processors in 1999 because of irreparable damage to the inter-processor network. [3]The air-cooled T3E system with 40 450-MHz processors, each with 128 MB of memory was replaced with a water-cooled T3E with 128 450-MHz processors, each with 256 MB of memory.
NOAA-NCEP	[1]IBM-SP [2]SGI/Origin 2000	[1]768 [2]256	[1]Nov. 1998; Major upgrade due in Sept. 2000 [2]Fall 1999
NPGS	[1]T3E [2]SGI Origin 2000 [3]IBM SP2 All off-site	[1]256 [2]128 [3]64	0–3 years old
NRL	[1]Cray C90 (2 systems at FNMOC) [2]Dec Alpha (NRL system) [3]SGI O2K (FNMOC) [4]T3E (DoD HPC/NAVO)	[1]16/8 [2]8 [3]128 [4]1088	[1]1999 [2]1999 [3]2000 [4]1998

Sustained System Performance	Central Memory / Secondary Disc Storage	Future Upgrades Planned
Sustained system performance of approximately 14–15 Gflops for the laboratory's actual workload.	Central Memory: [1]0.004 TB (Shared Memory) [2]0.001 TB (Shared Memory) [3]0.033 TB (Distributed Memory) Secondary Storage: [1]32 GB [2]2 GB [3]0 GB Rotating Disc Secondary Storage: [1]450 GB [2]770 GB [3]430 GB	Acquire a balanced high performance system to replace the current SGI/Cray systems. The first phase of this new system is expected to provide at least a three-to-four-fold increase in performance. The second phase, should deliver a substantial increase in performance over the phase-one system.
Unknown.	[1]256MB/node on 384 nodes, ~96GB total [2]128GB total	-The IBM-SP will be upgraded to 128 nodes (2048 PE) system in Sept. 2000 -Further upgrades to increase capacity in 2001. -NAVO MSRC will continue to increase its total capacity by installing new systems such as Sun server and IBM SP.
[1]10 Gflops [2]10 Gflops [3]5 Gflops	0.5–1.0 GB	The remote systems have plans in the works for upgrades of 2x to 5x in computing power.
[1]6.4/3.2 Gflops [2]2.0 Gflops [3]40 Gflops [4]50 Gflops	[1]8GB/3 TB [2]8GB/1 TB [3]256GB/3.7TB [4]387GB/1.5 TB	SGI O2K will be upgraded to SGI SN1 during fall 2000. DoD HPC undergoes constant upgrades.

TABLE 3-1   Continued

Institution[b]	Computer System	Processors	Last Upgrade
PNNL- S. Ghan	[1]~3 SUN ultra 5 workstations [2]Beowulf cluster	[1]1 [2]16	[1]1999 [2]2000
PNNL- R. Leung	[1]IBM-SP2	[1]512	[1]1999
PSU	[1]Cray SV-1 [2]IBM RS6000 SP (8 Winterhawk nodes)	[1]16 (each 1.2 GF) [2]8 nodes of 4 cpus each (32)	[1]2000 – Cray SV-1 replaced a J-class machine [2]Brand new

[a]This table does not include the potentially large amount of classified computer capabilities located at Department of Energy Laboratories that are occasionally used for climate modeling by in-house research groups.
[b]Institutions defined in Appendix J.
[c]No more than 1/16 of the total number of processors is applied to a single operation.

Sustained System Performance	Central Memory / Secondary Disc Storage	Future Upgrades Planned
[1]0.2 Gflop [2]2 Gflops	[1]512 Mb /30 GB [2]4 GB/320 GB	Upgrade Beowulf network to gigabit.
[1]247 Gflops	[1]262 GB/5 TB	Upgrade IBM-SP by replacing all existing processors with faster ones.
[1]6 Gflops [2]6 Gflops	[1]4 GB/220 GB [2]16 GB/292 GB	The IBM is an effort to match the architectures of recent U.S. lab purchases. If successful in transitioning codes to this machine the plan is to increase the number of cpus, hopefully by a factor of 3.

3.3 HUMAN RESOURCES

The survey responses revealed an overwhelming need at many of the modeling centers for highly qualified technical staff, such as modelers, hardware engineers, computer technologists, and programmers, who are difficult to find because private industry lures them away with higher salaries and other financial incentives.

An interesting point to note from the survey responses is that staffing levels at all three sizes of centers are similar despite differences in the scale of effort. This is likely because at the smaller centers many of those listed as staff are students and post-docs, whose number vary depending on funding levels. There are approximately 550 full-time employees dedicated to climate and weather modeling in the United States. This number is likely to be low because all small modeling centers were not surveyed, and there were a few intermediate-size centers that did not respond.

Most centers, regardless of size, indicated the likelihood of increasing the number of staff in the near future. Although many of the staffing increases listed were in the area of software development and computational support, a number of institutions were also increasing the scientific staff devoted to model interpretation and parameterization. Larger centers tended to be more satisfied with their staffing numbers. In part, this difference appears to be due to difficulties in finding stable, long-term funding for permanent staff at the small centers.

The respondents from universities differed in the belief that there is a decrease in the availability of high quality graduate students entering the atmospheric sciences. Those centers that felt there were sufficient students noted that the greater difficulty was finding continued funding to support the highest quality students available.

3.4 THE HIGHER-END CENTERS

Table 3-1 gives a synoptic view of the computer resources available to the higher-end centers in the United States. In general, most of the centers have computer capabilities on the order of 20 Gflops with one or two having twice that. With these resources most coupled climate models are run at about 300 km resolution in the atmosphere and about 100 km in the ocean.

In contrast, the European Center for Medium-range Weather Forecasting (ECMWF) has a 100-processor Fujitsu VPP5000 rated at a sustained 300 Gflops, a 116-processor Fujitsu VPP700 rated at a sustained 75 Gflops, and a 48-processor VPP700E rated at a sustained 34 Gflops. Its forecast model is run at 60 km resolution globally while its seasonal-to-interannual predictions are run at about 130 km resolution globally in a one-tiered sense and with ensembles of 15 per month. For more detailed information refer to *http://www.ecmwf.int/research/fc_by_computer.html*.

The Japanese Frontier Program is developing a 10 km global atmospheric model and has contracted for a supercomputer ("The Earth Simulator") having a sustained speed of 5 Tflops (*http://www.gaia.jaeri.go.jp/ OutlineOfGS40v3_1.pdf*).

3.5 ORGANIZATIONAL BACKGROUND

The earlier modeling report (NRC, 1998a) pointed out the basic health of small-scale climate modeling and the lagging progress of high-end climate modeling: these findings were confirmed above. That report summarized the difficulties faced by high-end climate modeling as follows: "The lack of national coordination and funding, and thus sustained interest, are substantial reasons why the United States is no longer in the lead in high-end climate modeling." It also identified the United States Global Change Research Program (USGCRP) as the only available mechanism to coordinate and balance the priorities established by individual agencies, but pointed out that the USGCRP did not have the means to do this.

More background is appropriate and again the organizational comparison of weather and climate proves valuable. The government organization for weather and weather forecasting was solidified about 1970 when NOAA and its Weather Service was placed in the Department of Commerce. The Weather Service embodied a specific agency structure with a well-defined mission that could be evaluated by progress in the production, accuracy, and delivery of weather forecast products.

The development of climate research in the United States was hastened by concerns over the perceived problem of global warming, but was constrained by the existence of an agency structure that had solidified by 1970. No additional government re-organizations occurred after 1970 and previous ones did not have climate as a tangible concern. Because no single agency could address all the aspects of climate (or more precisely, because many agencies claimed different aspects of climate but none were founded with climate as a mission), the Global Change Research Act of 1990 established the U.S. Global Change Research Program (USGCRP) "aimed at understanding and responding to global change, including the cumulative effects of human activities and natural processes on the environment, and to promote discussions toward protocols in global change research and for other purposes" (Appendix A of NRC, 1999a). It set into motion the USGCRP interagency process that addressed the following research elements:

1. global observations of "physical, chemical and biological processes in the earth system";

2. documentation of global change;
3. studies of earlier global change using paleo proxies;
4. predictions of global change including regional implications;
5. "focused research initiatives to understand the nature of and interactions among physical, chemical, biological, and social processes related to global change."

It also called upon the National Research Council to evaluate the science plan and provide priorities of future global change research. This was the motivation behind the NRC "Pathways" report (NRC, 1999a).

The Pathways report pointed out the flaws in the conception and implementation of the USGCRP—in particular that "in practice, the monitoring of climate variability is not currently an operational requirement of the USGCRP nor is there an agency of the U.S. government that accepts climate monitoring as an operational requirement or is committed to it as a goal." It also expanded the domain of climate research to include variability on seasonal-to-interannual and decadal-to-centennial time scales.

A group of agencies, each devoted only to research and combined in the USGCRP, is currently the only institutional arrangement for performing climate research; for establishing and sustaining a climate observing system; for identifying, developing and producing climate information products; for delivery of these products; and for building the general infrastructure needed to accomplish these tasks. The USGCRP is currently the only entity organized to develop climate models and to secure the computational and human infrastructure needed to respond to the demands placed on the climate modeling community. About 6% of the $1.8 billion annually allocated to the USGCRP is devoted to modeling and this includes the major data assimilation efforts of the NASA Data Assimilation Office.

3.6 SUMMARY OF HIGH-END CAPABILITIES IN THE UNITED STATES

With a sustained computer capability of 20 Gflops, the current capability of some of the U.S. high-end centers, a climate model consisting of a 300 km resolution atmosphere with 20 levels in the vertical, a land model, and 100 km ocean model, all coupled together and well coded for parallel machines is able to simulate 5–10 years per wall-clock day (see *http://www.cgd.ucar.edu/pcm/sc99/img002.jpg*. A 1000-year run would therefore take between 3 and 6 months to complete as a dedicated job. As we will see in the next section, these run times are too long to address some of the recent demands placed on the U.S. climate modeling community.

4

Increased Societal Demands on U.S. Modeling

There are clear societal needs and mandates to which the modeling community must respond. As these needs arise, the ability of the modeling community to develop and deliver relevant products as it is presently constituted continues to fall behind (NRC, 1998a). Understanding the influence of climate change on environmental cycles, both at a regional and global scale, is crucial to various aspects of society. For example, the combination of observations of climate change and projections of future alterations to climate resulting from anthropogenic inputs and natural variability have increased the awareness of the importance of accounting for the impacts of climate change. As a result, there have been increasing demands on the climate modeling community to provide climate data for use in assessments of the impacts of climate change at various time scales, on regional and global scales. This section outlines some of these recent demands.

4.1 OZONE ASSESSMENTS

The discovery of the catalytic destruction of ozone by chlorinated compounds, in particular chloroflourocarbons (CFCs), led to a number of international assessments of the physical and chemical states of the stratosphere under the auspices of the World Meteorological Organization (WMO) and the United Nations Environment Program (UNEP).

The first of these assessments (WMO, 1982) dealt with the observations and theory of ozone chemistry using one- and two-dimensional models of the chemistry of the stratosphere. In an assessment leading up to the Vienna Convention in 1985 (WMO, 1985) the situation had not changed very much although the use of full three-dimensional models for off-line tracer advection for use in chemical calculation was just becoming possible.

The Montreal Protocols, which limit the release of CFCs into the atmosphere, were signed in 1987, and the 1989 assessment (WMO, 1989) was input to these protocols. The state of modeling had advanced significantly with the availability of three-dimensional models of the stratosphere with the recognition that some stratospheric chemistry problems, such as the Antarctic ozone hole, were inherently three-dimensional. The Montreal Protocols required that the parties to the protocols to assess the control measures on the basis of available scientific, technical and economic information, at least every four years beginning in 1990. Notable in the first assessment (WMO, 1991) are chapters on the radiative forcing of climate, the role of ozone as a greenhouse gas and an evaluation of the effects of aircraft and rockets on the ozone layer. The expansion of interest of the assessment seems to have required rapid turnarounds so that mostly simplified models were used for the ozone chemistry which had since become extremely complex. Both the 1994 and 1998 Assessments (WMO, 1994; WMO, 1998) are dominated by two-dimensional models, but there were indications that fully three-dimensional models of the coupled climate and stratospheric chemistry were coming online.

The relationship between ozone and the climate problem, wherein ozone is affected by the climate of the atmosphere and in turn affects the climate, indicates the difficulty of the problem and points to the time when common tools will be used for both the Ozone Assessments and the Intergovernmental Panel on Climate Change (IPCC) assessments (below). As long as these remain separate, it appears that only two-dimensional models can, at present, address the full range of purely stratospheric chemistry concerns. As the problems are recognized to be intimately related and mutually dependent, it is expected that the demands on the modeling community for both assessments will converge.

4.2 IPCC

The IPCC was established in 1988 under the joint auspices of the WMO and the UNEP: "to assess the scientific, technical and socio-economic information relevant for the understanding of the risk of human-induced climate change. It does not carry out new research nor does it monitor climate-related data. It bases its assessment mainly on published and peer reviewed scientific technical literature" (http://www.ipcc.ch/about/about.htm).

The IPCC is organized into three scientific working groups. Only Working Group 1, which is devoted to scientific aspects of the global climate and its changes, concerns us here. Working Group 1 publishes an assessment every five years. The third assessment report has been recently released. Because the assessment reports are consensus documents involving hundreds of world scientists and because the Working Group Reports have international state review and endorsements, the assessment reports have demonstrated unusual authority and serve as standard sources for climate and climate change information.

The involvement of U.S. scientists in the IPCC process is through participation in writing the assessment reports, in reviewing the drafts of the report either as an individual scientist or as part of the governmental review, and perhaps most importantly, as producer of the scientific information on which the reports are based. The USGCRP program office supports individual scientist's participation in the IPCC writing process, organizes the U.S. governmental review, and provides the venue, resources, and scientific and technical personnel to support international Working Group II on Impacts, Adaptation, and Vulnerability.

Although participation by individual modelers in the IPCC Working Group I is informal and voluntary, it is a point of national pride that national models be included. Indeed, it is hard to see how the major industrialized nations of the world could make vital national decisions about greenhouse gases unless their own scientists have been involved. The Hadley Centre, for example, was created within the United Kingdom Meteorological Service to perform the data analyses and modeling runs needed to feed the IPCC process, especially with respect to detection and attribution of existing climate change and projection of future climate change. It is in these areas, which require long runs of coupled climate models, that the United States has had no similar concerted response, and has been shown to be lagging (NRC, 1998a).

4.3 U.S. NATIONAL ASSESSMENT

The USGCRP has been mandated by statute to undertake scientific assessments of the potential consequences of global change for the United States. The Global Change Research Act of 1990 (P.L. 101-606) states that the federal interagency committee for global change research of the National Science and Technology Council "shall prepare and submit to the President and the Congress an assessment which –

1. integrates, evaluates, and interprets the findings of the Program and discusses the scientific uncertainties associated with such findings;

2. analyzes the effects of global change on the natural environment, agriculture, energy production and use, land and water resources, transportation, human health and welfare, human social systems, and biological diversity; and

3. analyzes current trends in global change, both human-inducted and natural, and projects major trends for the subsequent 25 to 100 years."

This first U.S. national assessment (*http://www.gcrio.org/nationalassessment/*) includes a set of regional assessments, assessments of the consequences of climate change on five important societal and economic sectors of the nation (water resources and availability, agriculture and food production, human health, forests, and coastal areas), and a synthesis for policymakers.

The National Assessment used two model scenarios for long-term climate change: one produced by the Hadley Centre and one by the Canadian Climate Centre. These global models were chosen mostly because their output was available in time for the first step of the National Assessment, presentation of the scenarios to regional meetings so that the individual regions could study scenarios for future changes and possible responses in their own regions. The dependence of the National Assessment on foreign model results is contrary to the recommendations outlined in NRC (1998a), which argues that it is inappropriate for the United States to depend on foreign models for decisions about its own national interests. However, further review (Box 4-1) illustrates the difficulty of having a U.S. model respond to the needs of the National Assessment.

The 20 U.S. regions were asked to consider the two differing scenarios using the best guess of atmospheric concentrations of radiatively active constituents over the next 25–100 years. The regions were then asked to interpret the results and uncertainty of the results for their regions in terms of the interacting effects on such elements as water, energy, ecosystems, coasts (if any), forest, agriculture, and quality of life. The regional specificity of the two global models was poor with a resolution of T42 or approximately 300 km at the latitude of the United States. At the resolutions of these models orography was severely truncated, and it became difficult to assess future water resources in those regions that depend on mountain icepack for meltwater, since the extent of such icepack was badly misrepresented in the models and the height of the mountains was generally too low; the effect of warming on the icepacks was therefore generally too large (e.g. Plate 2).

The public law that called for the assessment requires a similar assessment every four years although the magnitude of the task was hardly foreseen by the authors of the law. A regular assessment would be conducted continually at specified intervals (probably no more frequently than 10 years or so) so that

1. The inter-communication that produces the assessment would continue.

Box 4-1
The National Assessment

In the winter of 1998 an official of the USGCRP asked scientists at the National Center for Atmospheric Research (NCAR) if they could run one or more simulations with the NCAR Climate System Model (CSM) that could be used in the National Assessment of Climate, which was being planned at that time. The planning committee for the assessment had decided to use data from two models, one from Canada and the other from the United Kingdom, as a starting point. The USGCRP official felt it would be desirable to have at least one American model being used for the assessment. There was a fairly tight time line to produce the data, approximately 10 months, in order for CSM data to be given to the people who would carry out the assessments.

The NCAR scientists recognized that this would be a major undertaking. They had no readily available emission scenarios for the twent-first century, and it was unclear how quickly credible scenarios could be developed. It was also unclear how much could be done with important components, particularly interactive sulfate aerosols, that were not included in the original CSM. A major complication was the lack of supercomputer time to carry out the necessary runs.

The Climate of the 20th Century run performed using the CSM took the equivalent of three months of fully dedicated Cray C-90 time. (This means all 16 processors of the C-90 running 24 hours a day, 7 days a week. Actually, the C-90 was never fully dedicated to the CSM because of the demands from competing modeling groups. The actual time for the run was closer to 6-12 months to completion.) The C-90 was part of NCAR's Climate Simulation Laboratory, which supported a variety of modeling activities. It was not possible, for the U.S. Assessment runs to use the fully dedicated machine. When it became clear that NCAR could not meet the deadline using the computers at NCAR, the USGCRP official volunteered to ask other agencies participating in USGCRP whether they had time available on a C-90 for this project. They did not.

NCAR scientists continued to work on the project, and two scenarios were developed; (1) a "business as usual" scenario with no political intervention to restrict greenhouse gas emissions; and (2) a "doubling carbon dioxide" scenario with interventions that restricted emissions to levels where the concentration of carbon dioxide in the atmosphere would asymptote to 550 ppm shortly after 2100. These scenarios were developed before any scenarios were developed for the most recent IPCC report, but they do not differ greatly from the IPCC scenarios. An interactive sulfate aerosol component model for the direct effect only was completed and tested in the atmosphere model. The scientific parts of the project were successfully carried out. Unfortunately, no extra computer time was found and the deadline was not met. NCAR completed the runs (using private resources to buy computer time in Japan) after the deadline, and data was made available to the assessment community, but it was not used.

---

2. Comparison of successive assessments might demonstrate a deeper insight into the response of the global climate and the regional climate.

3. The regions would demonstrate progress in understanding the integrated changes likely to occur and would gain a feeling for the vulnerability of their societies and institutions.

4. The questions arising from the ongoing assessment would stimulate the local regions' research agenda into their own modes of life and activities.

The ongoing nature of the assessment indicates that the demands on models will also be ongoing. The inability of the main U.S. modeling institution to come to grips with the first U.S. National Assessment (Box 4-1) and the inability of the United States to address climate change assessment requirements (NRC, 1998a) indicates that responding to the national assessment in the future will be a major problem for the U.S. modeling community.

4.4 SEASONAL-TO-INTERANNUAL FORECASTING

The development of seasonal-to-interannual forecasting grew out of the Tropical Ocean-Global Atmosphere (TOGA) program and the resulting understanding achieved in simulating and forecasting the El Niño/Southern Oscillation (ENSO). The history of TOGA and the development of short-term climate forecasting has been well documented in NRC (1996) and in a special issue of the Journal of Geophysical Research (Vol. 107, Issue C7, June 29, 1998).

The early realization of the usefulness of the forecasts led to the establishment of the ENSO Observing System in the tropical Pacific Ocean (NRC, 1994), which because of this perceived usefulness, survived the end of TOGA. It is being maintained as a quasi-operational observing system to this day (McPhaden et al, 1998). Because the ENSO Observing System provides the initial conditions needed to make forecasts of the phases of ENSO, a number of different seasonal-to-interannual forecasting efforts were established throughout the world, all using the data produced by the ENSO Observing System. Significant seasonal-to-interannual efforts in prediction exist at many places in the United States, in Australia, at the ECMWF, and in Germany.

A comparison of the prediction activities at the National Centers for Environmental Prediction (NCEP) and ECMWF provides insight into a potential path for seasonal-to-interannual prediction in the United States. (Ji et al, 1996; Barnston et al., 1994). The NCEP coupled model is a two-tiered system to save computer time: it first initializes and then predicts the tropical Pacific SST using a T40 (approximately 300 km) atmosphere coupled to an ocean in which only the tropical Pacific is active. The active Pacific has 150 km resolution in the zonal direction and 30-km to 100-km resolution in the meridional direction from the equator to 45°. The tropical Pacific SST is calculated monthly 6 months in advance. The global SST, with the predicted tropical SST, and the rest of the global SST started at observed values, and relaxed to climatology, is used to force a global atmospheric model at higher resolution to predict the global effects of tropical Pacific SST. An ensemble of 18 members of the T40 NCEP model is then used to predict the climate over the United States six months in advance. The final outlook, a part of the NCEP products suite (available at *http://www.ncep.noaa.gov/*) is subjectively determined from this ensemble and other statistical forecasts and is issued once a season.

In contrast, the ECMWF predictions are run in a one-tiered sense: the T63 atmosphere (about 200 km resolution) is coupled to a global ocean with resolution similar to NCEP, high at the equator and reducing gradually with latitude. The global ocean is initialized each week using all global data available, and an ensemble of seven 6-month predictions is run every week with the full coupled model.

Both NCEP and ECMWF release public forecasts of ocean SST. The NCEP outlook is for the United States only, while the ECMWF issues forecasts for the global tropics, Africa, South America, and East Asia; all other forecasts are available to member states of the European Centre only.

The potential economic value of these forecasts has been explored in detail (a bibliography is maintained at http://www.esig.ucar.edu/biblio/comprehensive.html); the public sector and many industries use these forecasts and clamor for more forecast skill. A large amount of research on seasonal-to-interannual predictability and prediction is being conducted (The July 2000 issue of the *Quarterly Journal of the Royal Meteorological Society* was devoted to the results of two programs, PROVOST (Prediction of Climate Variations on Seasonal-to-Interannual Time Scales) and DSP (Dynamical Seasonal Prediction), which are looking toward coordinated atmospheric GCM simulations in response to specified SST). A bibliography of several hundred references on seasonal-to-interannual forecasts is being maintained at <http://www.atmos.washington.edu/tpop/pop.htm>. It is clear that many nations, including the United States, will expand their research in seasonal-to-interannual predictions, create operational forecast systems, and apply the results for public good and private gain.

4.5 DECADAL AND LONGER VARIABILITY

One of the major advances of climate research over the last decade or so has been the realization that decadal and longer variability in the past has taken place in only a handful of patterns, in particular the Pacific Decadal Oscillation (PDO), the North American Oscillation (NAO) and its counterpart, the Arctic Oscillation (AO), the Atlantic Subtropical Dipole, the Antarctic Oscillation, and a number of other more regional patterns (NRC, 1998c). These patterns of variability have a profound effect on water resources, storms, food and fish resources, energy production and consumption, and the general economy and well-being of societies. Over the United States increasing amounts of variance of temperature and precipitation are successively explained by ENSO, the PDO, and the NAO (Higgins et al, 2000), implying that being able to predict these patterns would explain successively greater amounts of these crucial climatic variables, with obvious social and economic implications. The response of these patterns to the addition of radiatively active constituents to the atmosphere is also an active field of research with the idea gaining currency that global warming intensity and patterns cannot be understood without understanding the changes of these decadal patterns with time.

One of the recommendations of the IPCC Third Assessment Report was that patterns of long-term climate variability should be addressed more completely. This topic arises both in model calculations and in the climate system. In simulations the issue of climate drift in model calculations needs to be clarified in part because it compounds the difficulty of distinguishing signal and noise. With respect to the long-term natural variability in the climate system *per se*, it is important to understand this variability and to expand the emerging capability of predicting patterns of such organized variability as ENSO. This predictive capability is both a valuable test of model performance and a useful contribution to natural resource and economic management.

The possibility of projecting these patterns was discussed in NRC (1998c), and indications of predictability of the NAO (Rodwell et al, 1999; Saravanan et al, 2000), the PDO (Venzke et al, 2000), and the subtropical Atlantic Dipole (Chang et al, 1998) have been demonstrated. The actual and simulated forecasting of these patterns requires a tremendous amount of computer resources, comparable to ensembles of global warming simulations. We expect the study of decadal variations and the predictability of these variations to continue with the aim of discovering useful future predictability as a guide to long-term planning similar to the case of global warming simulations.

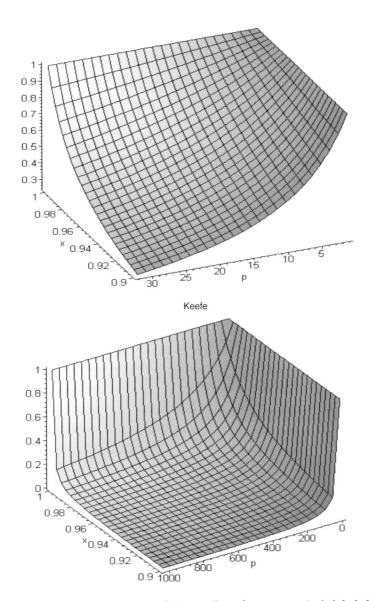

Keefe

PLATE 1 Amdahl's Law: Impact of the number of processors (axis labeled p) on the parallel efficiency (vertical axis) for a range of code parallelization (axis labeled x), between 90% and the ideal but unattainable 100%. a) Multiprocessing system with a number of processors between 1 and 32. Note that the greenish area, representing practically attainable efficiencies, ranges from over 20% for low parallelization to well over 50% for high parallelization. b) Massively parallel processor, with 40 to 1024 processors. Note that the attainable ranges of efficiency are well below 10% except for the hard-to-reach highest ranges of code parallelization (over 99%).

a.

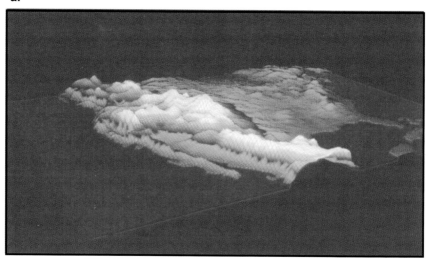

b.

PLATE 2 a. Terrain over North America on a model grid at 240 km horizontal resolution. b. Terrain over North America on a model grid at 30 km horizontal resolution. (Provided by M. Blackmon)

5

Responding to Climate Modeling Requirements

The following is needed in order to respond to the climate modeling requirements described in the previous sections:

1. additional computational resources, and human resources;
2. new mechanisms of standardization and interchange of models, model components, and model output;
3. increased availability of standardized diagnostic tools for standardized model evaluation;
4. new modes of organization and management.

5.1 COMPUTATIONAL RESOURCES REQUIRED

Model Resolution

A weather or climate model begins with fundamental equations governing the motions of the atmosphere, oceans, and sea ice, which are derived from physical laws, particularly the conservation of mass, momentum, and energy (e.g., Washington and Parkinson, 1986). These equations also include any other algorithms describing the movement of tracers and reactive chemicals through the modeled system. A number of recent books document the extensive treatments of various aspects of model building and the numerical methods needed to implement the models (Trenberth, 1992; Randall, 2000; Kantha and Clayson, 2000).

For the purposes of this analysis an atmospheric resolution of 30 km is used. This scale is viewed as a minimum resolution for acceptably modeling the orographic influences (Plate 2) and in particular to assure that a drop of water falling as precipitation falls into the correct catchment basin. It is also the minimum scale that can resolve weather motions characteristic of the climate being simulated. While it can be questioned whether such high resolution is needed for the climate simulation, it should be noted that the downscaling needed for applications is a one-way process. It uses the large-scale simulation as boundary conditions for smaller-scale simulations, but does not feed back on the larger scales. Thus, the highest resolution possible for the climatic simulation will assure the best possible downscaling. We will not argue that 30 km is the ultimate scale needed in the atmosphere but simply use it as a characteristic desirable scale needed to model the atmosphere for climate simulations. Such definiteness is also needed to estimate the computational resources required in this chapter. This resolution may be insufficient for many hydrological problems where resolution is important to adequately represent the groundwater paths and the many interacting spatial scales that affect water on its path from clouds to land to its ultimate return to the ocean. A land soil moisture and vegetation model must be embedded in the atmospheric model.

The relationship between horizontal and vertical scales is defined by the Rossby radius, such that a 30 km scale in the horizontal gives 300 m scales in the vertical. This implies a need for 50 layers up to the tropical tropopause at a height of 15 km. These layers may be augmented by increased resolution in the surface boundary layer. Including the stratosphere in the model requires additional vertical levels. Any sub-grid scale processes, including clouds, turbulent mixing, and boundary layer processes, not resolved by 30 km resolution, must be parameterized in terms of quantities on the resolved scale.

The ocean model is similarly coded from the equations of motion and the conservation of water and salt. To resolve western boundary currents a resolution of at least 10 km in the horizontal and 100 m in the vertical is required. The resolution in the vertical may be enhanced by extra layers to resolve the surface mixed layer. A sea ice model must be embedded in the ocean model in order to get the surface albedo of Earth correct and to ensure the correct salt balance of the ocean.

Because the ocean heat capacity is large and changes slow, it generally takes about 10,000 model years to spin up a coupled climate model to equilibrium, although acceleration techniques are available to speed the process. The model must be run at least 1,000 years, starting from near this equilibrium state, to diagnose its climatology and variability. Only after these diagnoses are performed is the climate model ready for use.

Ensembles

Because climate models may be sensitive to small changes in initial conditions, ensembles of many runs are made, each with slightly different initial conditions. The spread of the model results across the ensemble gives some idea of the sensitivity of the run to initial conditions and therefore the uncertainty of the final results.

To get an idea of the total output spread ensembles usually consist of 10-30 members. Practical exigencies of each problem determine the time allowed to complete the ensemble of runs. As a rough guide an ensemble of runs for weather prediction must be completed each day, for seasonal-to-interannual prediction each month, and for long-term climate change each year or two.

Computer Resources for the Various Types of Modeling

Simulation

Approximately 10^{17} floating-point operations are required to run a 30-km, 50-level atmospheric model for one model year. It takes roughly the same number of floating point operations to run a 10-km, 50-level ocean model for one year. Including all sub models and couplings, we can assume it takes of order $5*10^{17}$ floating point operations to run a coupled-climate model one model year.

A thousand-year model run would therefore take $5*10^{20}$ floating-point operations. As a rule of thumb, to develop a climate model, at least 10,000 simulated model years are required. Therefore $5*10^{21}$ floating point operations are needed to develop a model.

We see that on the order of 10-100 Tflops of *sustained* computer speed is needed to develop climate models at the specified resolution in reasonable amounts of time (Table 5-1). Adding components to form more comprehensive climate models takes still longer. Coupling additional chemical components of carbon models can increase these estimates by an order of magnitude. To analyze the variability and mean state of these runs a comparable amount of time is needed so that a number of thousand-year

---

Box 5-1
Capacity vs. Capability

Computing capacity — The ability to run many computing jobs, none of which requires all of even a large part of available computing resources. High computing capacity enables the throughput of many jobs that for climate modeling are often ensembles of runs with slight variations in initial or boundary conditions. Typically, these jobs can be run simultaneously thus providing a form of parallelism.
Computing capability — Simulations for which a single coupled or component model uses all or a large fraction of the entire system. Such calculations are much more demanding because they require the model code to efficiently use the number of processors available.

TABLE 5-1　Time to Perform a 10,000-Model Year Run at Various
Sustained Operating Speeds

Sustained Speed (Gflops)	Wall-clock Time (years)
5	30,000
100	1,500
1000	150
10,000	15
100,000	1.5

model runs can be done by a sustained 100-Tflops computer on the order of a year. Model development and climate simulation require the greatest capability.

Weather Prediction

Ensembles of 10-day forecasts with an atmosphere-only model are required. For the purpose of this exercise other crucial parts of a forecast system, such as quality control, data assimilation, and initialization, are ignored. At the resolution specified above, a ten-day forecast requires $2*10^{15}$ flop and an ensemble of 30 each day requires $6*10^{16}$ flops each day. Each member of the ensemble may be executed on a separate computer with the members potentially processed simultaneously. This requires the application of 20 Gflops acting over a day for each forecast, so a collection of 30 machines each operating at a sustained speed of 20 Gflops is required. This is an example of capacity computing since 20-Gflop machines are currently available. If the ensembles must, for some reason, be done sequentially each day, 600 sustained Gflops are required.

With the implementation of more sophisticated analysis schemes, such as four-dimensional variability (4-D VAR), the computational costs associated with preparation of the model initial states have become very significant. As an indication of this, of the ECMWF operational work, only 10% is attributed to the high-resolution 10-day forecast. The 4D-Var analysis consumes about 45% and the ensemble prediction system another 45%.

Short-Term Climate Prediction

For the purposes of this exercise, a six-month coupled forecast performed once a day so that we have an ensemble of 30 once a month is assumed. Again, the other parts of the forecast process are not included. Using the numbers for the resolution given above, $2.5*10^{17}$ flops need to be done each day which requires a sustained 2.5 Tflops.

Greenhouse Simulation

The resources needed to perform global warming projections are similar to those for simulations in developing and testing the models, except that an additional number of projections must be run. For *each* specified future concentration of radiatively active gases and constituents an ensemble of 100-year runs, each differing slightly in initial conditions, is needed.

Because the IPCC special report on emissions lists a number of future scenarios, an ensemble should be done for each emission scenario to simulate the full range of possible outcomes. Assuming a 10-member ensemble, 1,000 years of model years for each scenario or something on the order of 10,000 additional model years is implied. Thus, global warming projections put similar demands on high-end computing as simulation, and we can similarly conclude that a 10-100-Tflops computer would satisfy the needs for global warming projections. Additional computational demands arise when downscaling global warming projections to specific regions.

Compromises

In general it has not been the resolution needs of climate models that have determined computer purchases but rather computer availability that has determined the highest resolution at which climate models could be run. It was pointed out in Section 3.6 that at current resolution (about 300 km in the atmosphere) a thousand-year model runtime would take a current dedicated supercomputer running at 10–20 Gflops on the order of 3-6 wall-clock months to accomplish. Some Japanese computers can be bought *now* that run at a sustained 1 Tflops, a factor of 50 times faster, thus allowing a current increase in resolution of about a factor of 3 (100 km). The tradeoff between computer power and desired resolution is a compromise that will exist for a very long time.

5.2 WILL MASSIVELY PARALLEL ARCHITECTURES SATISFY OUR NEEDS?

Speed and Usability

The earlier modeling report (NRC, 1998a) detailed the NCAR procurement procedure that led to a Commerce Department anti-dumping tariff on Japanese vector computers. As noted in the discussion in Section 2.4, multi-processor machines are subject to limitations in speedup (Amdahl's law) such that the speedup factor over a single processor performance is less than the number of processors. The use of fast custom-designed processors helps to overcome the limitations of Amdahl's law.

For climate modeling, which has inefficiencies in code and a certain irreducible amount of sequential operations, parallel vector machines, currently built only by Japanese manufacturers, enjoy a throughput advantage of about a factor of 40 (per processor) over currently available massively parallel computers on real climate codes.

If Japanese supercomputers continue to be excluded from the U.S. market, we are faced with relatively limited options, all that remains are clusters of commodity-based SMP or PC nodes. The survey results (Appendix E) of the performance of current parallel climate models show that these machines do not compete effectively with Japanese VPPs in performance. Air-cooled Japanese vector machines are about 40 times faster (per processor) on climate and weather codes than current U.S. microprocessor-based machines. The latest Japanese vector machines are achieving impressive performance: 3–10 Gflops sustained per processor on real climate applications. The best microprocessors currently are achieving about 0.1 Gflops per processor.

These Japanese machines also have an additional advantage, that of *usability*. These machines have been developed in a sequence of incremental changes of the vector processor line that ended in the United States with the Cray T90. The software has developed slowly and carefully, and this generation of Japanese supercomputers, in particular the Fujitsu VPP line and the NEC SX-5 (*http://www.hpc.comp.nec.co.jp/sx-e/sx-world/*), have a full range of scheduling and balancing software and a robust compiler for FORTRAN. The machines are therefore usable, and the generation of coupled model codes written for vector machines work on them with minor modifications. In contrast, MPP and SMP machines are less mature and generally lacking the full range of software that allows immediate and facile use.

Vector or MPP or Both?

Massive parallelism and vector technology are not mutually exclusive. In fact, both Cray scalable vector machines and Japanese vector machines can be built with hundreds or thousands of processors. The Japanese have done a good job in building scalable interconnects for massively parallel vector machines in their NEC SX-5 and Fujitsu VPP5000: Both machines can be scaled up to 512 processors. The Earth Simulator project in Japan is building a network of parallel vector supercomputers with as many as 4096 processors designed to deliver sustained speeds of 5 Tflops.

Because this study also shows that much faster computers with larger memories are required to meet the needs of the U.S. climate modeling community, the choice is stark. If U.S. climate scientists are not able to purchase Japanese vector machines, they will continue to be unable to compete with their European and Asian counterparts.

Well-designed parallel applications can be written so that they scale and perform well on both vector and microprocessor-based machines. It is quite straightforward to do this for problems that use domain decomposition for parallelism; nearly all climate applications use domain decomposition. For cache-based microprocessors domains should be sub-divided into small two-dimensional patches so that better cache performance is achieved, while vector machines perform better with long vectors, so that long, thin patches are more appropriate. Note that either partitioning can be achieved with the same code. Only a few parameters need be changed to achieve either decomposition in the same code.

Nevertheless, massively parallel machines cannot provide infinitely scalable performance. Amdahl's Law is a stern taskmaster, such that massively parallel processors require that the code be almost perfectly parallel to obtain good performance, which is difficult to achieve. Synchronization, load imbalance, and serial execution inherent in some codes makes achieving 99% parallelism very difficult, yet a code that is 99% parallel can achieve at most 100-fold speedup, even on a thousand-processor machine (Plate 1).

Most environments for climate could use both types of machines. Vector machines for mainframe performance, scalability, and reliability, and particularly for capability computing as defined in Box 5-1. SMP and MPP machines are well suited to capacity computing applications requiring large throughput of multiple jobs and for pre- and post-processing of data sets, and for data assimilation, where the ratio of vector to microprocessor speeds is low (between 4 and 6).

The Bottom Line

Clearly the easiest path to good performance is for U.S. scientists to have access to machines with the most powerful processors, the smallest and fastest network, and the fastest memory access, making cache unnecessary. That the Japanese computers are currently superior in both speed and usability for weather and climate model codes is without question. It is no coincidence that in countries other than the United States, the great majority of weather services and climate research institutions have purchased, or are about to purchase, Japanese parallel vector computers. We note finally that a previous NRC report (NRC, 1998a) has stated: "The United States must apply greater resources, particularly (but not exclusively) in the area of advanced computer machines. National boundaries should not influence where machines are purchased."

5.3 THE NEED FOR CENTRALIZED FACILITIES AND OPERATIONS

Viewed purely as research the 20+ intermediate-size and large U.S. climate-modeling activities are healthy and diverse, allowing many people to engage in productive and innovative climate research. Viewed from the point of being able to bring concentrated resources to bear on specific problems, the 20+ differing climate activities could be considered duplicative, inefficient, and sub-critical. This shortcoming is illustrated by one of the responses to the survey, which stated that "the first problem in the United States is fragmentation, and very inefficient use of resources// single agencies, such as NOAA and NASA are unable to organize an integrated global modeling effort within their agency." Owing to the increasing range of climate products required by diverse users, there is increasing need for a single agency charged with assembling and disseminating the various climate products to these diverse user groups.

One solution to this problem would be to have a few centralized climate modeling activities under the auspices of a single agency, which are of critical size and have adequate resources, each devoted to a specific task. These centralized modeling activities would maintain close linkages to the various research and user groups and would undertake model building, quality control and validation of models and products, product design, regular and systematic product development, and integration of observational data. Although these operational activities would be centralized, they would take advantage of research activities external to operations, including model development, analysis, diagnostics, and interpretation.

Perhaps the most telling argument for centralization of operational climate modeling is by analogy. Every major country in the world that invests in weather services has chosen to have the forecasts centralized in a modeling and prediction center, usually co-located with other weather activities.

5.4 FOSTERING COOPERATION WITH A COMMON MODELING INFRASTRUCTURE

The Efficiency of Cooperation

Small modelers in the United States have modern workstations available to them, sometimes with more than one processor. These workstations can be used for coarse-resolution ocean models run for tens of model years; high-resolution atmospheric models for a few days or weeks; development of radiation codes, boundary layer models in the atmosphere and ocean, and new numerical schemes; and the diagnosis and analyses of observed data and the output of large models.

By definition high-end modelers have supercomputers available to them. In the United States, this group is unique because it can currently run high-resolution ocean models for hundreds of model years, atmospheric general circulation models for hundreds of model years, very-high-resolution global atmospheric models for a few model months, coupled-atmosphere ocean models for hundreds of model years, ensembles of seasonal-to-interannual forecasts (in two-tiered configurations) for six months to a year in advance, and ensembles of weather forecasts. Until recently these two groups of modelers have tended not to interact.

The desirability for interaction of these classes of modelers is overwhelming and brings benefits to both classes. Small-scale modelers who develop parameterizations prove the worth of these parameterizations in a climate model and to demonstrate this worth they need access to high-end computers and models. High-end modelers gain the expertise of a large number of smaller modelers. The output of large models is usually underanalyzed and making it available to a wider community not only gives smaller modelers and diagnosticians access to a system approximating the climate so that they may understand its mechanisms but also gives invaluable feedback to the high-end modelers about the fidelity and faults of their climate models.

An effective U.S. modeling effort therefore requires better cooperation. This in turn requires an extensive and effective shared infrastructure that facilitates the exchange of technology and provides means and metrics to rigorously benchmark, validate, and evaluate models and forecast systems. It should, for example, allow university researchers access to an integrated forecast system, through which they can investigate the impacts of a new process parameterization. It should facilitate and support the interaction between focused centers and the broader research community for development and process experimentation, as well as evaluation and diagnostics of simulations and predictions.

The technical difficulties of exchange can be satisfied by the common modeling infrastructure (CMI), but the full potential of interaction can be realized only if access to computers and the free availability of models and model output is assured on a mutually beneficial basis.

Common Modeling Infrastructure

The process of developing, evaluating, and exercising complex models and model components is resource intensive. It places serious demands on personnel, computing, data storage, and data access that cannot be met by any single group or institution. Incorporating new understanding or new technology into comprehensive models and forecast systems additionally requires effective collaboration and communication between the research community and model developers. Currently, the U.S. modeling effort is limited in both of these respects. The distribution of effort is less of a problem than the fragmentation of efforts, most operating with their own standards for codes, data, and model and forecast evaluation. The technical effort required to port codes and data from one center to another or to incorporate new components into an existing system often prevents effective interaction, and even communication between groups. As a result, the enterprise as a whole is inefficient, and progress is slowed.

The concept of CMI was initiated at the NSF/NCEP Workshop on Global Weather and Climate Models (NSF/NCEP, 1998), where the participants agreed that global atmospheric model development and application for climate and weather in the United States should be based on a common model infrastructure. The CMI was proposed to address the growing perception that the diversity of models currently in use at U.S. modeling centers is acting as a barrier to collaboration among groups. To this end a CMI steering committee was established to develop a flexible modeling infrastructure with standards and guiding principles to facilitate the exchange of technology between operational and research, weather and climate modeling groups in the United States. The proposed goals were to accelerate progress in global numerical weather prediction (NWP) and climate prediction, to provide a focal point and shared infrastructure for model development, and to provide a means for assessing physical parameterization schemes. A CMI does not imply a common model; it is simply the set of standards and protocols that ensures common features of different models and compatibility of files not only among different models but also with observational data.

Once established the CMI group recommended the creation of core models devoted to a particular modeling focus (e.g., numerical weather prediction, seasonal-to-interannual prediction, decadal variability). In the development of these core models it was recommended that each should concentrate on a problem that would benefit from broad involvement of the modeling community and that should be associated with a center whose mission is directly related to that problem (such as NCEP for NWP). Furthermore, the core models should be based on a flexible common model infrastructure and permit a range of options for different physical problems with standard configurations defined for operational applications. These configurations would represent the primary development path for the core models and would provide controls upon which improvements could be tested. Candidates for core models would be based on well-defined code standards designed to advance the goals of the common model infrastructure. New codes should be straightforward to enable the integration of new diagnostics to model output and to implement new parameterizations. Core model codes should also be portable to several machines.

Software Framework

The key to making models and model components interoperable is the existence of a common framework that is flexible, modular, and efficient. The framework needs to separate the scientific and computational components of the code in order to facilitate the separate development efforts in each area and to ensure the greatest portability across multiple computing architectures. Framework prototypes have been developed at a few centers, and there is consensus on the outlines of a broader-based framework even at present. Development of a community framework, however, represents a significant software development exercise over a period of several years, with continuous support for maintenance and evolution thereafter.

Data Standards

Models and analysis tools interface with data constantly. Researchers collaborate extensively through data exchange. The research enterprise will be aided greatly by standards that minimize the effort in accessing and exchanging information. Perhaps most important is the development of community metadata standards and conventions. With standardized metadata, applications programs and models can be tailored to exchange, access, and manipulate data efficiently and effectively, even in a remote site, with minimal effort. This ultimately will lead to greater interoperability between software applications and easier interaction between individuals and between groups.

Community Modeling Repository

A comprehensive and well-supported community modeling resource will allow the full potential of established coding frameworks and data standards to be realized. The repository will house a full array of model components and physical parameterization codes, including full dynamical cores and integrated physics packages of one or more "standard" models (supported by large development and support efforts in the national program), all fully documented, with validation output, and fully consistent with the community modeling framework and standards. It will allow a graduate student at any university access to complete coupled or uncoupled models and to components that could be used to construct a new model for research. The advantage of the repository for modeling centers will be the ease with which model components and parameterizations can be exchanged with other groups and the ease with which their model can be ported, tested, and run at numerous other sites. To be effective the repository must be well staffed in software engineering, information management, and quality control. The staff will actively solicit continuing expansions and updates of the repository, ensure compliance with standards, and support the use of repository resources throughout the community.

Several other common needs can also be effectively met through a repository including statistical analysis, diagnostic and visualization tools, standard algorithms (such as model clocks and calendars), physical constants, and certain standard data (e.g., vegetation, surface elevation). The repository will simplify the locating of possible tools or resources for model-based development and research. It will include adequate documentation to inform users how to access and use the resources and will be fully compatible with the community modeling framework, making access, porting, and use of resources straightforward.

Common Tests and Evaluations

Development inevitably involves testing and evaluation. Though much of this work is specific to particular research projects, there are important benchmark simulations and results that are of widespread scientific (and societal) interest. Examples include climate change simulations with prescribed trace gases (e.g., IPCC-related simulations) and seasonal forecast suites. A valuable component of an enhanced common modeling infrastructure is the establishment and maintenance of community benchmark calculations and evaluation metrics. These would fit easily into a general repository structure, providing continually clear and succinct results with which state-of-the-art assessments can be made and evaluation of new tools measured.

5.5 HUMAN RESOURCES

In the surveys referred to in Chapter 3, the respondents commented that, after the availability of computer time, their most pressing need was for computer technologists to convert code from existing machines to the new parallel machines and in general to optimize code for use on this new class of machines. At a time when Internet companies offer large salaries or stock options or both, it becomes very hard for research grants to compete for computer technologists. Even though some technologists want to stay in a research environment despite the relatively small monetary rewards, the numbers are small, and research groups compete for the few that are available. This problem is one of the unintended consequences of the enforced shift from vector to massively parallel machines available to the U.S. high-end modeling community.

Indeed, most U.S. Earth science centers are experiencing increased turnover in computational positions, with a net migration away from the field. Significant numbers of earth scientists are leaving the field after school, rather than moving into scientific positions. To compete with the non-scientific information technology job market, scientific organizations need to offer not simply competitive salaries but also development of job skills that are attractive to mainstream professionals and career paths comparable to others in the scientific field. There is also a disturbing tendency for decreases at the front end of the employment pipeline, with 1998 showing a 20% drop in the number of Ph.D.s awarded in meteorology and oceanography over 1997 (*UCAR Quarterly*, Spring 2000), which was not mirrored in other scientific fields. According to *UCAR Quarterly* this drop in graduates is mirrored by a drop in enrollments: "During the past year, members of the UCAR Board of Trustees have expressed concern about a perceived sudden drop in the number of qualified students applying to their graduate departments."

The staffing issues discussed above are likely to continue their negative impact on climate modeling science into the near future. To prevent the continued drain of competent scientists and technicians to overseas institutions and to the information technology (IT) sector requires such actions as increasing salaries at modeling institutions and improving career development for technical staff.

One technical solution to this problem can be modeled after the Application Service Provider (ASP) phenomenon. An ASP is a commercial company that creates a shared service center accessible over the Internet. Companies pay monthly fees to the ASP for the shared services of a limited number of IT professionals. Using this concept in a centralized computer facility would allow climate scientists to have access to the small number of IT professionals available to the entire community. The overall human resources issue is so fluid and so deeply rooted in the economic and social conditions in the United States that, aside from noting the problem and its likely effect on any attack on the climate modeling problem, it is difficult to present any global solutions.

5.6 NEED FOR CLIMATE SERVICES AND MANAGEMENT ISSUES

The current approach of expecting existing organizations within the USGCRP to deliver climate information products as an activity ancillary to their primary missions has not been successful (NRC, 1998a). Simply providing these organizations with small amounts of additional funding to give them incrementally greater capability is therefore not an effective remedy to the current situation. To provide the required capabilities for climate modeling activities and to insure the production of climate modeling products, there needs to be some organizational entity with its primary mission being the delivery of these products. For the sake of this discussion only, this entity will be designated a Climate Service—no agency or other organizational connotation is implied. The discussion then will delineate the properties of this Climate Service needed to deliver climate information products.

Institutional and Incentive Issues

The Climate Service must have a clearly defined mission, focused on the delivery of the product and the assurance of product quality. These products should be the result of the scientific process, though the delivery of the product will require bringing closure to incomplete scientific arguments to allow production of software suites. Different versions of the model generating climate products need to be tested and validated prior to their use for product generation.

The defined mission of the Climate Service will be to provide an overarching structure to facilitate prioritization of institutional needs and decision making. Though the mission is essential, it is also important that there is an executive decision-making function vested in a small group of science and software managers, whose performance is measured by the successful delivery of the products and subsequent customer response. At the lead of this group will be an individual with the ultimate authority and responsibility for product delivery.

The current fragmented situation does not support an effective incentive structure at any level. At the lowest level scientists are generally rewarded for individual accomplishments of discovery-driven research. At the next level, even in the most project-focused organizations, funds flow into organizations from a variety of program managers. The program managers naturally command the allegiance of these subsets of the organization and are generally not rewarded for the delivery of successful products by the organizations they fund. This programmatic fracturing extends to computational resources, and in most U.S. laboratories there is a disconnect between computational resources and the delivery of simulation and assimilation products. The disconnect arises because the computing organization is often funded to pursue computational research in information technology programs or the computing facility is run as an institutional facility and the product generation exists in an uncomfortable balance with large numbers of small discovery-driven research projects. Finally, the organizations that are expected to deliver the needed Earth science products are often embedded in large Agency laboratories whose basic metrics of success do not include delivery of successful Earth science simulation and assimilation products. All told, the current structure of Earth science activities in the United States is fracturing rather than unifying.

For an executive function to be effective, an organization has to have an incentive structure that connects all facets of the organization to the responsibility for successful product generation.

Business Practices

A functioning Climate Service that contains the attributes described above would stand in stark contrast to the pervasive scientific culture of the United States. Such an organization would vest the decision-making function in an executive process that acts in the best interest of the delivery of the institutional products. Such a Climate Service will require supporting business practices that are significantly different from those currently used in the scientific community. These business practices must be unifying. They must provide a mechanism for stable and effective external review and integration with the discovery-driven research community.

As with the scientific and computational aspects of this enterprise, the business practices need to be considered in a systematic and integrated way. They need to support the goals and function of the charged institute. While the complete specification of these business practices are beyond the scope of this document, the following can be derived from experience in the current organizations.

Funding should be:

1. focused on delivery of products;
2. stable over 10-year time periods;
3. balanced on all elements of the organization;
4. under the direction of the executive decision-making function responsible for scientific quality and operational success;
5. isolated from the program volatility of funding agencies.

Review:

1. conventional peer review will not work;
2. need to develop review techniques to support organization, including review of science and operations;
3. different levels of review are needed for scientific and operational purposes.

Business practices:

1. success of the Climate Service must be a critical metric for success of the hosting agencies;
2. contractual vehicles must support the organizational goals;
3. salary structures must allow effective recruiting and retention of personnel.

5.7 REWARDING THE TRANSITION WITHIN THE RESEARCH COMMUNITY

There must be an incentive for the research community to develop societally useful products for transition to the Climate Service. The situation can be explained most easily by considering the transition from research to operations. Suppose the research community develops something valuable, such as the development of seasonal-to-interannual prediction and the ENSO Observing System (see NRC, 1996). Research from the TOGA program demonstrated the predictability of aspects of ENSO. From this an observing system was established, and insight into the kind of questions that must be answered in order to use these types of climate forecasts were asked (NRC, 1999c).

A natural transition would then be to recognize the value of ENSO predictions, and on the basis of demonstrated value, move the prediction aspect and the routine observations needed to initialize the prediction into the operational domain using new resources in anticipation of demonstrated benefits. Instead, most of the prediction and all the observing system has remained in the research domain. Resources that should be used to explore and develop new knowledge is therefore going into activities that are not research but undoubtedly contribute to research. As a result, the "reward" for the research community to develop seasonal-to-interannual prediction has been decreased financial resources. As noted in the Pathways report (NRC, 1999a), "A research program can maintain a permanent observing system only when the system is relatively cheap and does not inhibit other research objectives. When there is an operational need for a system, funding must not come from research sources, else the building of a permanent observing system could gradually impoverish the research enterprise."

Developing societally valuable research that leads to climate information products should lead to a clear transition path whereby the products find a home in the Climate Service. The reward to the research community should be the freeing up of resources so that research can address new problems, perhaps leading to new societal benefits.

5.8 PROVIDING THE BEST POSSIBLE SERVICE TO AN INFORMED PUBLIC

A Climate Service focused on the production and delivery of climate information must make these products as useful as possible to its customers. Weather forecasting has dealt with similar problems for a long time and therefore provides a framework for modeling for societal benefit; some of the discussion is based on lessons from that arena. To provide the best weather and climate services, effective interaction with informed customers is essential. To do this requires meeting several challenges.

The first challenge is to make sure that the most current and reliable information reaches the public. The products must be authoritative and one way of assuring this is to have an unbiased organ of the governmental bureaucracy either produce or bless the product. Professional organizations, such as the American Meteorological Society (AMS) and the American Geophysical Union (AGU), strive to put out the best information possible and they are answerable to professionals in the field. The public makes the final choice of what to believe, however; so the educational system must emphasize the basics of science and critical thinking to the generations of future voters.

The second challenge is to engage the scientific community in such efforts. Scientific institutions often state education as a goal, but lack of commitment at the supervisor level or by peers and narrowly defined reward structures can discourage scientists from engaging in public or educational outreach activities. Such commitment involves allowing time and resources for training and the outreach activities themselves.

The third challenge is to decrease public confusion about climate issues. Public exposure to climate change is often in the form of sound bites explaining the latest weather disaster in terms of El Niño or global warming. Yet climate issues are difficult to understand without going back to the basics. Trying to explain the question of natural versus anthropogenic climate change to the public, for example, involves many issues, including:

1. how climate change is measured over different time scales (issues of pollen proxies, sea life, crop records, and more recently, instrument corrections);

2. what determines climate and climate change (changes in greenhouse gases, aerosols, land-surface properties, solar output, ocean);

3. the physical processes (especially radiation);

4. what a numerical model is;

5. how a climate model is tested (against past and present climate, testing of parameterization schemes against special data sets, studying the way the model responds to data input if run as a weather model); and

6. what the differences in climate models really mean.

Clearly, teaching such material involves not a single lecture but a carefully crafted set of activities and discussions that the audience (typically teachers) can use to pass on the information. Because climate and weather sciences evolve, a means of getting new information (e.g., Web sites) is included, along with contact information for future questions.

The fourth challenge is maintaining strong links between the forecasting and user communities and their customers. All sides must agree on what is needed, what is reliable, what is most usable, and what is realistic. This is best met when the first three challenges are met. *It is essential, however, that the providers learn from the customers, or that they learn together.* The very nature of the modeling products produced by the Climate Service must be negotiated between the service and its customers. This involves not only formal interaction but also research on societal aspects of use of weather and climate information (Pielke and Kimpel, 1997). We expect that the creation and distribution of useful climate products for the public and private use will be the best way of maintaining these links.

5.9 SUMMARY

Increased computational and human resources are required to effectively respond to the various demands outlined in Section 4. A new way of focusing resources to meet the specific challenges posed by these various demands implies a less fragmented and therefore more centralized mode of addressing these problems. The nature of the institutional and management requirements were discussed in terms of a Climate Service, which here is the designation for the organizational entity that would create the climate information products and manage the climate modeling activities that would deliver these products. The full range of functional components of such a Climate Service extend beyond climate modeling and were not discussed. This will be presented in Section 7, where an overall vision of its functions and its interaction with the research community is presented.

6

Improving the Effectiveness of U.S. Climate Modeling

Currently, U.S. climate modeling is characterized by a highly creative, productive, and healthy community, particularly at the intermediate and smaller scales of effort. Despite this success, high-end modeling capabilities lag behind those of other nations. The Panel on Improving the Effectiveness of U.S. Climate Modeling was convened to provide federal agencies with an assessment of the nation's technical modeling needs and to provide recommendations on how government, interacting with the scientific community, can optimize the use of modeling talents in the United States.

The two primary users of climate models are the research community, whose goals are to advance the understanding of the climate system, and the operational community, which uses models for the production of climate prediction products in response to societal demands. The two groups are connected because research is crucial for constructing and evaluating the models needed to produce useful products, and the operational infrastructure required to produce these products (such as a sustained observing system, modeling system, and resulting model assimilated data products that result) are of great utility to the research community.

There are, however, differences between these two groups. The research community's approach to modeling is exploratory and sometimes without a clear path, with success judged by peer review. The operational community operates with a more clearly defined mission and success is often judged by the utility of a given product. Operational modeling is more rigidly constrained by external controls and evaluation procedures requiring a dependable infrastructure with a high level of organization and centralization.

During its deliberations, the panel identified the key issues influencing the effectiveness of U.S. climate modeling. The following sections identify these issues and outline the panel's recommendations. These recommendations are based on input from a survey distributed to many of the U.S. modeling centers, a workshop held in Washington, D.C., and the expertise of the panel and address some of the missing elements in U.S. climate modeling.

The Need for Centralized Operations

Information about future climate is crucial for addressing numerous societal needs. Different communities and subsets of society require distinct climate-related products. Thus, centralized modeling activities under the auspices of a single agency are needed to assemble and distribute the necessary climate information and products to diverse user groups. The United States has not as yet centralized its climate activities.

Centralized modeling activities should have close linkages to research and user groups and ought to include model building, quality control and validation of models and products, product design and regular and systematic product production, and integration of observational data. Centralized activities require computational systems adequate to address these problems. Although the operational activities would be centralized, they should take advantage of research activities external to operations, including model development and analysis, diagnostics, and interpretation.

The European Centre for Medium-Range Weather Forecasts (ECMWF) is one model of a successful dedicated modeling facility. The center was established as a European cooperative weather forecasting venture. During its lifetime it has produced some of the highest-quality, highest-resolution forecasts from any modeling group. It is not clear, however, that this model translates into leadership in modeling research or research into the long-term aspects of the climate system. Although this model may be successful in Europe, it is not clear that it can be applied to the more decentralized U.S. climate modeling community.

Many of the measurements and observations used to define climate are made in the arena of weather prediction, and many of the atmospheric processes and feedbacks that influence short-term weather contribute to climate. Therefore, the panel recognizes that strong weather forecasting capabilities are necessary preconditions for effective climate model development, and close ties should be maintained between climate and weather modeling activities.

Finding: Increased demands for operational climate products of benefit to society, such as those required for the IPCC and National Assessments and for short-range climate forecasts, have placed heavy demands on the research community, which is neither well suited by culture nor by organization to regularly produce these products. Despite this the research community is essential in providing the knowledge needed to develop effective climate products.

Finding: When comparing U.S. and European high-end modeling the panel finds that U.S. modeling is still lagging in its ability to rapidly produce accurate high-resolution model runs. This situation has worsened since the publication of the NRC report *Capacity of U.S. Climate Modeling to Support Climate Change Assessment Activities* (NRC, 1998a).

Recommendation 1: In order to augment and improve the effectiveness of the U.S. climate modeling effort so that it can respond to societal needs, the panel recommends that enhanced and stable resources be focused on dedicated and centralized operational activities capable of addressing each of the following societally important activities:

1. short-term climate prediction on scales of months to years;
2. study of climate variability and predictability on decadal-to-centennial time scales;
3. national and international assessments of anthropogenic climate change;
4. national and international ozone assessments; and
5. assessment of the regional impacts of climatic change.

The Need for Open Access to the Most Appropriate Computer Architecture

The most effective means to obtain superior computer performance is to utilize machines with powerful processors; small, fast networks; and rapid memory access. Currently, Japanese parallel vector supercomputers provide the fastest and most capable architecture for the sustained processing of climate model codes. Access to these computing systems for U.S. scientists is limited due to the high tariffs put on these systems as a result of a Commerce Department anti-dumping order.

In order to provide high-quality climate products as well as operate an effective research program, the climate modeling community should have access to state-of-the-art, high-end computing facilities connected with centralized modeling activities as discussed previously. These computing facilities should have sufficient capabilities to comprehensively investigate the climate system and to enhance and develop models to better understand climate change. To provide the most capable computational resources available, national boundaries should not influence where computers are purchased (NRC, 1998a).

Finding: Current U.S. high-end modeling efforts are being hindered by the forced acceptance of a computational architecture ill-suited to process the algorithms employed to model the earth's complicated climate system.

Recommendation 2: The panel recommends the adoption of a scientific computing policy ensuring open access to systems best suited to the needs of the climate modeling community.

Recommendation 3: Researchers should have improved access to modern, high-end computing facilities connected with the centralized operational activities discussed in Recommendation 1. These facilities should be sufficiently capable to enable comprehensive study of the climate system and help develop models and techniques to address relevant high-end climate modeling problems.

The Need for a Common Modeling Infrastructure

Effective climate model development is often hindered by an incompatibility between model components developed in different groups. A protocol to facilitate the scientific exchange of common diagnostic tools, the interchange of model components, and the exchange of data in a common data format is not yet in general use. To aid the evaluation and exchange of technological and research advances within and among the research and operational modeling communities, a set of common diagnostic and visualization tools and a set of programming and data standards is needed. The panel believes that a set of common modeling tools and standards would enhance the cooperation between high-end modelers and smaller scales of effort, would increase the efficiency of climate model development, and would reduce duplication of efforts among groups.

Finding: A common modeling infrastructure, consisting of system software and model code would reduce the inefficiencies within the climate modeling research communities and would allow the research and operational communities to interact successfully.

Recommendation 4: In order to maximize the effectiveness of different operational climate modeling efforts, these efforts should be linked to each other and to the research community by a common modeling and data infrastructure. Furthermore, operational modeling should maintain links to the latest advances in computer science and information technology.

Human Resource Needs in Support of Climate Modeling Activities

The climate modeling community is facing an overwhelming shortage of qualified technical and scientific staff. This difficulty is, in part, due to the inability of both research and operational modeling centers to compete with the high salaries and incentives offered by the high tech industry. Some overseas groups, (e.g. ECMWF) have overcome this difficulty by providing highly lucrative salary packages that modeling groups in the United States are unable to match. This situation is aggravated in U.S. university-based modeling groups since they are often dependent on the vagaries of short term funding for employee salaries. Furthermore, this situation is impacting university graduate programs as many students receive lucrative offers from private industry prior to the completion of their degree. This human resource problem is reflected by declining graduate enrollments in all areas of the climate sciences and in the growing disparity in the quality of life of scientists, especially young ones, living in major cities, and their private sector counterparts.

Finding: The panel finds that there is currently a strain on human resources in the climate modeling community. U.S. modeling groups are having difficulty competing with private industry and with overseas institutions for the high skilled and experienced scientists and computer technologists needed to ensure an effective modeling effort in both research and operational modeling efforts.

The shortage of highly skilled technical workers is not particular to the climate modeling community, but is part of a larger shortage of human resources affecting nearly all areas of science and engineering. The complexity of this problem, and the lack of expertise on the panel to address this issue, precludes this panel from making any specific recommendations related to human resources.

Institutional Arrangements for Delivery of Climate Services

The panel has argued that the suite of designated "Climate Services" consisting of the establishment and sustenance of a climate observing system, the production of useful model products on the global and regional scale, and delivery and dissemination of these useful products would satisfy societal demands and would be of great benefit to the research community. In what form these climate services would be delivered; how much of the climate information would be developed and delivered by public versus private sources; which agencies would take part; what roles they would play; and how the links between operational modeling, sustained observations, and research are to be established and maintained are all open questions and beyond the expertise of the panel.

Recommendation 5: Research studies on the socio-economic aspects of climate and climate modeling should be undertaken at appropriate institutions to design the institutional and governmental structures required to provide effective climate services. This assessment should include:

1. an examination of present and future societal needs for climate information;

2. a diagnosis of existing institutional capabilities for providing climate services;

3. an analysis of institutional and governmental constraints for sustaining a climate observing system, modeling the climate system, communicating with the research community, and delivering useful climate information;

4. an analysis of the human resources available and needed to accomplish the above tasks;

5. an analysis of costs and required solutions to remove the constraints in accomplishing the above tasks;

6. recommendations on the most effective form of institutional and governmental organization to produce and deliver climate information for the public and private sectors.

7

A Vision for the Future

The panel believes that successful implementation of the recommendations given in the previous chapter would go a long way toward making both modeling for societal benefit and modeling purely for understanding much more effective. The panel also notes that the integration of operations and research for both modeling and observations is a prerequisite to successfully producing and delivering useful climate information. The panel is compelled to augment its view that addressing modeling alone is not enough, and that climate modeling can be made more effective by the creation of operational entities that maintain observing systems, produce and disseminate climate information, and are responsible for coordinating certain functions with the climate research community and the weather forecasting operational community.

High-end climate modeling depends on observations and on research. Likewise, climate research depends on high-end modeling and observations. Observational data assimilated into a comprehensive coupled climate model enables the verification and enhancement of model code to produce accurate and continual climate analysis. From panel deliberations and acknowledgements of the robust linkages between research, observations and climate modeling, the panel endorses previous NRC reports (NRC, 1998b, 1999a, 1999b, 1999c, 2000a, 2000b) that called for the development of a sustained climate observing system. In this section, the panel presents a vision of an operational entity that facilitates the synergy among sustained climate observations, high-end modeling, and research while identifying, creating, and producing climate information useful to society.

7.1 CLIMATE RESEARCH AND CLIMATE OPERATIONS

We have distinguished modeling for societal benefit and modeling for understanding. Modeling for societal benefit is product oriented, requires regular and systematic runs of climate models, and is user driven and user evaluated. Modeling for understanding is freer, competitive, driven and evaluated by scientific priorities, evaluated by peers, and generally less capable of being organized or constrained. We recognize that these distinctions are not universal and are therefore imperfect, but we have found it useful to make these distinctions in order to consider the needed organizational aspects for each type of modeling.

It is the nature of all climate research that a proper balance between process studies, background observations, and modeling best advances the understanding of the entire system. It is the difficulty of climate research that, while it is easy to put together field programs of limited duration to measure poorly understood processes, it is almost impossible to sustain measurements on climatic time scales. The climate research community neither has the infrastructure for doing so nor are sustained observations amenable to the usual peer review process, because sustaining observations is not itself research.

Climate operations has an analogous structure. It needs an observing system and it produces model products using high-end modeling, some for analyzing and improving the observing system and some for diagnostic and predictive information. A prime function of climate operations is the design and delivery of climate information products that benefit society and put demands on the observational system and on modeling. These societal functions are qualitatively different from research functions. They tend to have time constraints determined by the nature of the decision to be made, they require specific products to be delivered in forms most useful to decision makers, and they are judged by a different standard from curiosity-driven research. In practice the difference between this type of product-driven research and curiosity-driven research shows up as differences in resources and organizations required, which, in turn, implies different modes of management and funding. Product-driven research tends to be large scale (beyond the scale of a single principal investigator), more expensive, and more highly centralized. The additional resources required are justified in terms of benefits to users with the ultimate evaluation done not by modelers but by the users themselves. In particular, operations can (and must) sustain infrastructure and can (and must) sustain an observing system.

7.2 MUTUAL INTERACTIONS AND MUTUAL BENEFITS BETWEEN CLIMATE RESEARCH AND CLIMATE OPERATIONS

Operations provide enormous benefits to research and are most likely to be successful when interacting strongly with research. For example, the current weather observations network is maintained by the Weather Service for construction of weather forecast products. The long records of upper air observations, taken globally since the 1950s, are an invaluable resource for atmospheric researchers. Indeed, it is hard to imagine the upper-air network being maintained in the research domain in a principal investigator mode of operation. It is also hard to conceive of weather operations existing without the development and improvement of numerical weather forecasting, which arose in the research domain. It is this synergy between the weather forecasting operations and atmospheric research that is most valuable both to researchers and to society.

But we note the asymmetry between the two types of activities. Centralized, expensive, and ongoing operations can contribute greatly to curiosity-driven research, but research that is decentralized and organized predominately in a principal investigator mode cannot produce the extensive regular and systematic products demanded by society. It can, however, design and help develop these products.

It is this asymmetry between research and operations that must be recognized for us to present our vision. We envision a modeling activity that responds and contributes to *both* research and societal requirements. Because of the noted asymmetry, each function can be fulfilled only with the involvement of the other. Because the analysis of the state of the climate system also involves model assimilated data and because model development requires a constant confrontation between observations and models, we can diagram the needed interactions as in Figure 7-1.

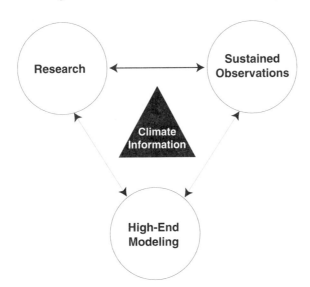

FIGURE 7-1 Interactions between climate information, high-end modeling, research, and sustained observations.

As in a similar model for research (Figure 2-2), the *interactions* and *balance* of components of Figure 7-1 are crucial. Research, in the usual distributed principal investigator mode, is needed to define and understand the climate problem and constantly improve the components of models through deepened understanding derived from process studies, small-scale models, and the diagnosis and analysis of observations. Sustained observations are the basis for our knowledge about the climate system. It involves long-term, accurate, and calibrated measurements of all components of the climate system. This element may use existing observations of the atmosphere, ocean, land, hydrosphere, etc. taken for different purposes but must assure they are maintained as climate observations (NRC, 1999b). The high-end modeling component synthesizes the research and assimilates the sustained observations to provide climate information products for use by researchers and society.

It is the concept behind Figure 7-1 that forms our larger vision of effective climate modeling in the United States. Modeling is thus put in its proper context: as synthesizer of research, as assimilator of observations, and as producer of products for society. It is in the implementation of this concept that the details of our vision must be fleshed out in the relationships implied by the arrows and in the organization and resources needed for the components. It is clear that by considering large-scale modeling and its proper context we have once again come upon the needed functional components of a Climate Service.

Within the conceptual picture of Figure 7-1, we begin with the high-end modeling component. The magnitude of the problem is large because it involves modeling the climate system and assimilating large amounts of data, including satellite data. As long as it involves any of the several activities detailed in Chapter 4, the climate models involved will be large, as close to comprehensive as possible, requiring simulations of thousands of model years to develop and many hundreds to apply.

An effective modeling activity for both operations and research should have models and model components developed and improved in the research domain, run in the operations domain, and analyzed in both domains. The process should be ongoing and cyclic and should involve focused parts of the research community in every phase of the cycle. It should have centralized parts in order to accomplish its operational mandate and may have a degree of decentralization for the research functions (consistent with need to run on high-end computers). This model should avoid the duplication and disorganization endemic (and probably necessary) to a successful small-scale research enterprise.

We see the optimal way of fulfilling this requirement as a small number of operational Centers (either new or existing, colocated or otherwise in touch), each devoted to a different societal need (i.e., producing a different product), each adequate in resources to its task, interacting with each other and the research community through a mechanism of exchange involving a common modeling infrastructure.

In addition to the development and operation of a high-end climate model, each center should include a facility (people, storage devices, and machines) for the storage and distribution of model codes and output to the research community and for the collection of parameterizations and diagnostics from the research community, all in common formats. Common diagnostic tools would be developed and used by all large modeling centers and by the researchers interacting with them. Research funding and Center computer time should be made available to the research community for model improvement. The Centers themselves would have a certain amount of funding available for the research they deem necessary for advancing their tasks.

To develop and evaluate the needed models, the Centers and the distributed research community must cooperate. The Centers build and run the models. They make available to the research community the output of the models. While universities do not have the resources to run the models, they can certainly diagnose their output. The university research community thus has access to the latest model output and the Centers gain an ongoing diagnosis of the state of the models.

Similarly, in the building of the models the parameterization of unresolved processes (e.g., clouds, mixing in the ocean, soil moisture) in the Center models should be competed for by interested researchers. The research community would gain the funding and resources (monetary and computer) for better defining climate processes, and the Centers would gain the expertise and perhaps manpower to run the extensive tests needed to find out if new parameterizations improve the model. One of the absolutely necessary functions of the Centers' interaction with the research community is enabling and facilitating the arduous process that takes researchers from the analysis of data from the synoptic network or from field programs to the development of improved parameterization schemes for use in climate models. Again, crucial to this interchange is the standardization of protocols of data and codes common both to the Centers and to the research community. This would also guard against duplication because the codes would be available to and used by all modeling Centers and all members of the research community.

Observations are critical for defining the state of the models and for providing the analyses against which the models can be tested. The climate models of course need to be tested against an analysis provided by a model different from the one being tested. Because much of the model atmospheric data is ingested for purposes of weather prediction at NCEP, it is important that the operational climate centers be tightly connected to NCEP.

The output of the Centers' analysis models should be made available to the research community. This should include ocean analyses, land analyses, coupled-climate analyses, and atmospheric reanalysis. The research community will benefit from having the various analyses to diagnose, and the operational climate observational community (when they exist) will benefit from the diagnoses. The Centers will have improvements in observing design for their own diverse purposes as one of their tasks, so that both observations and modeling will benefit from an optimized observing system.

We do not mean to imply that the entire climate research community should be engaged with climate operations—this would be neither practical nor desirable. But the benefits to be gained from having climate research interacting with climate operations would stimulate research and enrich operations to an extent that the benefits of interaction would be hard to overlook.

7.3 FROM VISION TO REALITY

Climate Research and Climate Operations are not interchangeable and both are needed to construct and disseminate climate information products for the benefit of society. Climate Operations will be expensive, with the major cost being the climate observing system. Because of the integrated nature of the functions needed for Climate Operations, high-end modeling must be considered an essential part of operations.

The nation needs the best possible climate information on which to base decisions about the future. The panel has no doubt that the nation will, at some point in the future, choose to institute Climate Operations. An effective high-end climate modeling activity is an essential step on the way.

References

Bailey, D.H., E. Barszcz, J.T. Barton, D.S. Browning, R.L. Carter, L. Dagum, R. A. Fatoohi, P.O. Frederickson, T.A. Lasinski, R.S. Schreiber, H.D. Simon, V. Venkatakrishnan, and S.K. Weeratunga. 1991. The NAS Parallel Benchmarks. *Int. J. Supercomputer Applic.* 5:66–73.

Barnston, A.G., A. Leetmaa, V.E. Kousky, R.E. Livezey, E.A. O'Lenic, H. Van den Dool, J. Wagner, and D.A. Unger, 1999: NCEP Forecasts of the El Niño of 1997–98 and its U.S. Impacts, *Bull. Am. Met. Soc.*, 80, 1829–1852.

Barnston, A.G., H.M. van den Dool, S.E. Zebiak, T.P. Barnett, M. Ji, D.R. Roedenhuis, M.A. Cane, A. Leetmaa, N.E. Graham, C.R. Ropelewski, V.E. Kousky, E.A. O'Lenic, and R.E. Livezey, 1994: Long-lead Seasonal Forecasts – Where Do We Stand? *Bull. Amer. Meteorol. Soc.*, 75, 2097–2114.

Bryan, K. 1984. Accelerating the Convergence to Equilibrium of Ocean-climate Models. *J. Phys. Ocean.*, 14:666–673.

Chang, P., C. Penland, L. Ji, H. Li and L. Matrasova, 1998: Predicting Decadal Sea Surface Temperature Variability in the Tropical Atlantic Ocean, *Geophys. Res. Let.* 25, 1193–1196.

Chang, P., L. Ji, H. Li, C. Penland, and L. Matrisova, 1998. Prediction of Tropical Atlantic Sea Surface Temperature, *Geophys. Res. Let.*, 25, 1193–1196.

Department of Energy, National Science Foundation. 1998. National Workshop on Advanced Scientific Computing: *http://www.sc.doe.gov/ssi/LangerRep.html*.

Department of Energy. 1998. The Accelerated Climate Prediction Initiative: Bringing the Promise of Simulation to the Challenge of Climate Change.: *http://www.epm.ornl.gov/ACPI/*.

Hennessy, J.L., and D.A. Patterson. 1990. Computer Architecture: A Quantitative Approach. Morgan Kaufmann Publishers, Inc.

Higgins, R.W., A. Leetmaa, Y. Xue, and A. Barnston, 2000: Dominant Factors Influencing the Seasonal Predictability of U.S. Precipitation and Surface Air Temperature. *J. Climate*, 13, 3994–4107.

Hollingsworth, A., M. Capaldo, and A.J. Simmons. 1999. The Scientific and Technical Foundation of the ECMWF Strategy 1999–2008. *ECMWF*.

Hooke, W.H. and R.A. Pielke Jr., 2000. Short-term Weather Prediction: An Orchestra in Need of a Conductor, in Prediction: Science, Decision Making and the Future of Nature, D. Sarewitz, R.A. Pielke, Jr. and R. Byerly, Jr. (eds), Island Press, Washington D.C. pp.61–84.

JASON, 1998. Letter to Dr. Ari Patrinos about ACPI. http://www.epm.ornl.gov/ACPI/Documents/jason_letter_report.html

Ji, M., A. Leetmaa, and V.E. Kousky, 1996: Coupled Model Forecasts of ENSO During the 1980's and 1990's at the National Meteorological Center. J. Climate, 9, 3105–3120.

Ji, M., D. W. Behringer, and A. Leetmaa, 1996: An Improved Coupled Model for ENSO Prediction and Implications for Ocean Initialization. Part II: The Coupled Model. Mon. Wea. Rev., 126, 1022–1034.

Kantha, L.H., and C.A. Clayson. 2000. Numerical Models of Oceans and Oceanic Processes. Academic Press, International Geophysics Series Vol 66, 940pp.

McPhaden, M. J., A.J. Busalacchi, R. Cheney, J.R. Donguy, K.S. Gage, D. Halpern, M. Ji, P. Julian, G. Meyers, G.T. Mitchum, P.P. Niiler, J. Picaut, R.W. Reynolds, N. Smith, and K. Takeuchi, 1998: The Tropical Ocean — Global Atmosphere Observing System: A Decade of Progress. J. Geophys, Res., 103, 14169–14240.

NRC, 1994: GOALS (Global Ocean-Atmosphere-Land System) for Predicting Seasonal-to-Interannual Climate: A Program of Observation, Modeling, and Analysis. National Academy Press, 116 pages.

NRC, 1996: Learning to Predict Climate Variations Associated with El Niño and the Southern Oscillation: Accomplishments and Legacies of the TOGA Program. National Academy Press, 192 pages.

NRC, 1998a: Capacity of U.S. Climate Modeling to Support Climate Change Assessment Activities. National Academy Press, 78 pages.

NRC, 1998b: The Atmospheric Sciences Entering the Twenty-First Century. National Academy Press, 364 pages.

NRC, 1998c: Decade-to-Century-Scale Climate Variability and Change: A Science Strategy. National Academy Press, 160 pages.

NRC, 1999a: Global Environmental Change: Research Pathways for the Next Decade. National Academy Press, 616 pages.

NRC, 1999b: Adequacy of Climate Observing Systems. National Academy Press, 66 pages.

NRC, 1999c: Making Climate Forecasts Matter. Paul C. Stern and William E. Easterling, Editors. National Academy Press, 192 pages.

NRC, 2000a: From Research to Operations in Weather Satellites and Numerical Weather Prediction. Crossing the Valley of Death. National Academy Press, 80 pages.

NRC, 2000b. Our Common Journey. National Academy Press, 363 pages.

Nordhaus, W.D, and J. Boyer. 2000. Warming the World: Economic Models of Global Warming. Cambridge, Mass.: MIT Press, 232 pages.

NSF/NCEP, 1998: Workshop on "Global Weather and Climate Modeling" http://nsipp.gsfc. nasa.gov/infra/report.final.html

Oliker, L., and R. Biswas. 2000. Parallelization of a Dynamic Unstructured Application Using Three Leading Paradigms. IEEE Transactions on Parallel and Distributed Systems.

Pielke, R.A., Jr., and J. Kimpel. 1997. Societal Aspects of Weather: Report of the Sixth Prospectus Development Team of the U.S. Weather Research Program to NOAA and NSF. Bull. Amer. Meteor. Soc. 78:867–76.

President's Information Technology Advisory Committee (PITAC). 1999. Information Technology Research: Investing In Our Future. http://www.ccic.gov/ac/report/

Randall, D.A., editor, 2000: General Circulation Model Development: Past, Present, Future. Academic Press, International Geophysics Series Vol 70, 807 pages.

Rodwell, M. J., D. P. Rowell, and C. K. Folland, 1999: Oceanic Forcing of the Wintertime North Atlantic Oscillation and European Climate. Nature, 398, 320–323.

Saravanan, R., G. Danabasoglu, S. C. Doney, and J. C. McWilliams, 1999: Decadal Variability and Predictability in the Midlatitude Ocean-atmosphere System. *J. Climate,* 13, 1073–1097.

Trenberth, K. A., editor:, 1992 Climate System Modeling,. Cambridge University Press

Venzke, S. M. Munnich, and M. Latif, 2000: On the Predictability of Decadal Changes in the North Pacific. *Climate Dynamics* 16, 379–392.

USGCRP, 2000. High-end Climate Science: Development of Modeling and Related Computing Capabilities.

WMO, 1981: The Stratosphere 1981 Theory and Measurements. WMO Global Ozone Research and Monitoring Report No. 11.

WMO, 1985: Atmospheric Ozone 1985. 3 vol. WMO Global Ozone Research and Monitoring Report, No. 16.

WMO, 1989: Scientific Assessment of Stratospheric Ozone:1989. 2 vol. WMO Global Ozone Research and Monitoring Report, No. 20.

WMO, 1991: Scientific Assessment of Ozone Depletion: 1991.WMO Global Ozone Research and Monitoring Report No. 25.

WMO, 1994: Scientific Assessment of Ozone Depletion: 1994.WMO Global Ozone Research and Monitoring Report No. 37.

WMO, 1998: Scientific Assessment of Ozone Depletion: 1998. WMO Global Ozone Research and Monitoring Report No. 44.

APPENDIXES

Appendix A

Steering Committee and Staff Biographies

Dr. Edward S. Sarachik (Chair) is a professor in the Department of Atmospheric Sciences and an adjunct professor in the School of Oceanography at the University of Washington. Dr. Sarachik's research interests focus on large-scale atmosphere-ocean interactions, seasonal variations in the tropical oceans, the role of the ocean in climate change, and biogeochemical cycles in the global ocean. He is vice-chair of the NRC's Climate Research Committee (CRC) of the Board on Atmospheric Sciences and Climate (BASC), was chair of the Tropical Ocean/Global Atmosphere (TOGA) Advisory Panel, and has been a member of numerous other NRC committees.

Dr. Lennart Bengtsson is director of Theoretical Climate Modeling at Max-Planck-Institut für Meteorologie, Hamburg, Germany. His research focuses on the use of coupled atmosphere/ocean general circulation models for investigating natural and anthropogenic climate change.

Dr. Maurice L. Blackmon is head of the National Center for Atmospheric Research's (NCAR) Climate and Global Dynamics Division (CGD), and was previously the director of the National Oceanic and Atmospheric Administration's (NOAA) Climate Diagnostics Center. He was a key player in the development of the first generation of the NCAR Community Climate Model (CCM). Dr. Blackmon also participated in climate diagnostic studies on the El Niño-Southern Oscillation and other phenomena. Dr. Blackmon has been a member of numerous NRC committees and is currently a member of the CRC.

Dr. Margaret A. LeMone is a senior scientist in the Mesoscale and Microscale Meteorology Division at NCAR. Dr. LeMone's research has focused both on the structure and dynamics of the atmospheric boundary layer and its interaction with the land surface and clouds and on the interaction of mesoscale convective systems with the boundary layer and surrounding atmosphere. Although primarily involved in the interpretation and synthesis of observations, she integrates results from field experiments with numerical studies through collaborations to better understand the physics and to improve the models. Dr. LeMone is a member of the National Academy of Engineering and a former member of BASC.

Dr. Robert C. Malone is employed at the Advanced Computing Laboratory at Los Alamos National Laboratory. Dr. Malone has many years of experience in numerical modeling of physical processes and systems, including stellar evolution, laser-produced plasmas, magnetically confined plasmas, Earth's atmosphere, and, most recently, Earth's oceans. He contributed to the development and validation of the first version of NCAR's Community Climate Model (CCM), then led a small team at Los Alamos to extend and apply the model for studies of the "nuclear winter" hypothesis.

Dr. Matthew T. O'Keefe is an associate professor at the University of Minnesota. His research is centered on parallel processing, with an emphasis on parallel computer architectures and compilation for these machines. Dr. O'Keefe formed the Parallel and Computer Systems Laboratory at the University of Minnesota, where research is both industrially and academically oriented. He has worked with others on the Miami Isopycnic Coordinate Ocean Model (MICOM) where his role was developing versions of MICOM suitable for execution on massively parallel processors and shared-memory multiprocessors.

Dr. Richard B. Rood is a senior scientist in the Data Assimilation Office (DAO) at NASA/Goddard Space Flight Center. Dr. Rood's scientific background is in the modeling of tracer transport and chemistry in the atmosphere, and more recently, climate modeling. In a previous capacity as head of DAO, he was involved in expanding the scope of data assimilation from numerical weather prediction applications to more generalized Earth science. In 1995 he received the NASA Exceptional Scientific Achievement Medal.

Dr. Stephen Zebiak is director of Modeling and Prediction Research at the International Research Institute for Climate Prediction at Lamont-Doherty Earth Observatory. Dr. Zebiak has worked in the area of ocean-atmosphere interaction and climate variability since completing his Ph.D. in 1984. He was an author of the first dynamical model used to successfully predict El Niño. He has served on various advisory committees, including the NRC TOGA Advisory Panel. He is presently co-chair of the U.S. CLIVAR Seasonal-to-Interannual Modeling and Prediction Panel, and is chair of the International CLIVAR Working Group on Seasonal-to-Interannual Prediction.

STAFF

Dr. Vaughan C. Turekian is a Program Officer with the Board on Atmospheric Sciences and Climate. He received his Ph.D. in Environmental Sciences from the University of Virginia in 2000. Dr. Turekian's research involves using stable bulk and compound-specific isotope analyses to characterize the sources and processing of aerosols in marine air. He has also used radiogenic isotopes to study residence times in the atmospheric boundary layer and in estuaries.

Dr. Alexandra R. Isern is a Program Officer with the Board on Atmospheric Sciences and Climate. She received her Ph.D. in Marine Geology from the Swiss Federal Institute of Technology in 1993. Dr. Isern was a lecturer in Oceanography and Geology at the University of Sydney, Australia from 1994 to 1999. Her research focuses on the influences of paleoclimate and sea level variability on ancient reefs. Dr. Isern is co-chief scientist for Ocean Drilling Program Leg 194 that will investigate the magnitudes of ancient sea level change.

Mr. Carter W. Ford is a Project Assistant for the Board on Atmospheric Sciences and Climate. He has been involved in a wide variety of NRC projects including studies pertaining to climate modeling, the GEWEX program, and the World Climate Research Programme. Prior to BASC, Mr. Ford served with the NRC's National Weather Service Modernization Committee. He holds a B.A. in International Studies from Miami University (Ohio).

Appendix B

Capacity of U.S. Climate Modeling to Support Climate Change Assessment Activities

EXECUTIVE SUMMARY

The U.S. government has pending before it the ratification of the Kyoto Protocol, an agreement to limit the emissions of greenhouse gases (GHGs), which is largely based on the threat GHGs pose to the global climate. Such an agreement would have significant economic and national security implications, and therefore any national policy decisions regarding this issue should rely in part on the best possible suite of scenarios from climate models.

The U.S. climate modeling research community is a world leader in intermediate and smaller[1] climate modeling efforts—research that has been instrumental in improving the understanding of specific components of the climate system. Somewhat in contrast, the United States has been less prominent in producing high-end climate modeling results, which have been featured in recent international assessments of the impacts of climate change. The fact that U.S. contributions of these state-of-the-art results have been relatively sparse has prompted a number of prominent climate researchers to question the current organization and support of climate modeling research in the United States, and has led, ultimately, to this report.

---

[1]An example of what is referred to in this document as a small modeling effort is one using a global, stand-alone atmospheric climate model at R15 (~4.5° × 7.5°) resolution; an example of an intermediate effort is one using a global, stand-alone atmospheric climate model at T42 (2.8° × 2.8°) resolution; an example of a large or high-end modeling effort is one using a global, coupled T42 atmospheric/2° × 2° oceanic model (or finer resolution) for centennial-scale simulations of transient climate change.

In this evaluation of U.S. climate modeling efforts, the Climate Research Committee (CRC) was asked by USGCRP agency program managers to address three key questions, which form the basis for the NRC Statement of Task (Appendix B) for this report:

1. Do USGCRP agencies have a coordinated approach for prioritizing from a national perspective their climate modeling research and assessment efforts?

2. Are resources allocated effectively to address such priorities? A related question that the report addresses is whether currently available resources in the United States are adequate for the purpose of high-end climate modeling.

3. How can the U.S. climate modeling community make more efficient use of its available resources?

- Regarding the first question — **the CRC has reached the conclusion that although individual federal agencies may have established well-defined priorities for climate modeling research, there is no integrated national strategy** designed to encourage climate modeling that specifically addresses, for example, the objectives of the USGCRP, the needs for comprehensive contributions to the IPCC science base, and the priorities developed by the CRC in its chapter in the Board on Atmospheric Sciences and Climate's report, *Atmospheric Sciences Entering the Twenty-First Century* (NRC, 1998a). We suggest that the science-driven climate modeling agenda, which has been largely shaped by individual investigators, has been reasonably effective in advancing the frontiers of science, but has not been adequately responsive to the immediate needs of the broader community (e.g., the "impacts" and "policy" communities).

- With respect to the second question — **we find that compared with intermediate and smaller modeling efforts, insufficient human and computational resources are being devoted to high-end, computer intensive, comprehensive modeling, perhaps, in part, because of the absence of a nationally-coordinated modeling strategy.** Consequently, in contrast to some of the foreign modeling centers, U.S. modeling centers have found it difficult to perform coupled atmosphere–ocean climate change scenario simulations at the spatial resolutions relevant to certain national policy decisions (e.g., finer than 500 km × 800 km). The recognized strengths of U.S. intermediate modeling capabilities (see, e.g., the sizable contributions from the U.S. coarse-resolution climate modeling efforts in the IPCC reports) have not been effectively harnessed in the development of high-end, U.S.-based models. For instance, leading Earth system modeling efforts in the United States suffer from a computationally limited ability to test and run models in a timely fashion. The ability of the climate community to acquire state-of-the-art mainframes is severely hampered by a Department of Commerce "antidumping order" prescribing a financial penalty in excess of 400 percent for the purchase of the world's most powerful commercial supercomputers, which are Japanese in origin. The climate community has not been provided with the financial or computational resources to overcome this barrier and has, therefore, been unable to fully capitalize on the scientific potential within the United States. Not only is insufficient access to powerful computers hampering scientific progress in understanding fundamental climate processes, it is also limiting the ability to perform simulations of direct relevance to policy decisions related to human influences on climate. However, *at least as important* as the insufficiency of computing resources are the lack of national coordination and insufficient funding of human resources.

- Regarding the third question — the CRC finds that:
  1. A national set of goals and objectives that are agreed to by the USGCRP agencies is essential.

  2. A concerted effort by the relevant agencies is needed to establish a coordinated national strategy for climate modeling.

  3. In order to optimally use existing scientific capabilities, adequate resources, including greatly improved supercomputing capabilities, need to be provided to the climate modeling community.

  4. The reliance of the United States upon other countries for high-end climate modeling must be redressed.

In order to avoid the aforementioned problem regarding priority setting, the USGCRP could assume increased responsibility for identifying, from an interagency perspective, any gaps or imbalances in the research priorities established by the individual agencies. At present, however, this is made difficult because some agencies have excluded from their USGCRP budgets the computational and human resources to support comprehensive, coupled atmosphere-ocean climate modeling efforts on par with those in several foreign countries. **Although an entirely top-down management approach for climate modeling is viewed as undesirable, national economic and security interests nevertheless require a more comprehensive national strategy for setting priorities, and improving and applying climate models.** An effective national approach to climate modeling should ensure that available resources are allocated appropriately according to agreed upon science research and societal priorities and are efficiently utilized by the modeling community. We acknowledge that justification for and design of such a strategy would require a more complete evaluation of the current status of climate modeling in the United States than was possible in developing this report. Development of such a strategy should take place with full involvement of climate modelers within academia and the national climate research centers, along with users of climate modeling results and agency program managers.

Climate modeling in the U.S. promotes a healthy competition among various groups, but without better coordination of research among national laboratories and between them and the academic community, it may be difficult to optimally utilize available human and high-end computer resources. In particular, standardization of model output, model evaluation tools, and modular programming structures can facilitate model development and minimize duplication of effort, with the possibility that prudent standardization may yield some cost savings. High-end modeling coordination could also be enhanced through refereed workshops to discuss the pertinent scientific and associated societal issues and to recommend priorities. Effective collaborative linkages between process studies and modeling groups should also be encouraged to facilitate the difficult task of developing, implementing, and testing new model parameterizations. In addition, increased coordination of research-based and operational modeling activities will help ensure that expertise in these two communities is shared. These are but a few of the types of coordinating activities that should be vigorously and consistently pursued.

The CRC finds that the United States lags behind other countries in its ability to model long-term climate change. Those deficiencies limit the ability of the United States:

1. to predict future climate states and thus:
    a) assess the national and international value and impact of climate change;
    b) formulate policies that will be consistent with national objectives and be compatible with global commitments;
2. to most effectively advance understanding of the underlying scientific issues pertaining to climate variability and change.

Although collaboration and free and open information and data exchange with foreign modeling centers are critical, it is inappropriate for the United States to rely heavily upon foreign centers to provide high-end modeling capabilities. There are a number of reasons for this, including the following:

1. U.S. scientists do not necessarily have full, open, and timely access to output from European models, particularly as the commercial value of these predictions and scenarios increases in the future.[2]

---

[2] U.S. researchers do, however, currently have access to output from most simulations of transient climate change produced by foreign models.

2. Decisions that might substantially affect the U.S. economy might be made based upon considerations of simulations (e.g., nested-grid runs) produced by countries with different priorities than those of the United States.

3. If U.S. scientists lose involvement in high-end modeling activities, they may miss opportunities to gain valuable insights into the underlying processes that are critical to subsequent modeling investigations. In this regard the issue of accessibility is much more than just a commercial and political issue; in order to most effectively advance the science in the United States, researchers need to have access to both model output and the models themselves to iteratively diagnose the output, advance our knowledge of climate, and improve the models' predictive capabilities.

4. There are currently relatively few modeling centers anywhere in the world capable of producing moderate resolution (e.g., 250–300 km grid spacing), transient climate simulations. The differences in simulated climate produced by each of these models' different structures help to bound the range of outcomes that the climate system might produce given a certain forcing scenario. Thus, the state of climate modeling throughout the world is such that the addition or removal of even a single model could affect the confidence levels assigned to certain scenarios of future climate change. In other words, not only would the United States benefit from enhancements in its modeling capabilities, the international community would benefit from these efforts as well. The marginal benefits from only modestly increased investments in comprehensive models in the United States could be very large, because, if properly coordinated, the enhanced emphasis on high-end modeling could be built upon the excellent existing U.S. strength in small and intermediate modeling.

Thus, to facilitate future climate assessments, climate treaty negotiations, and our understanding and predictions of climate, it is appropriate to develop a national climate modeling strategy that includes the provision of adequate computational and human resources and is integrated across agencies.

Appendix C

Questionnaire Sent to Large and Intermediate Modeling Centers

I. General information

1. Name of Institution:

2. Name of Research Group (if applicable):

3. Name and contact details of person completing this survey:
   Name
   Position
   Address

   Phone
   Fax
   e-mail

4. Please indicate the primary funding sources for your modeling efforts and an approximate percent breakdown where there is more than one source.

5. How would you describe the purpose(s) of your modeling efforts?

6. What percentage of your modeling activities are devoted to operational versus research purposes?

7. Please describe groups with which you have significant collaboration and briefly describe the nature of this collaboration.

II. Current U.S. Modeling capabilities

1. Please provide your opinion on current U.S. climate and weather modeling capabilities relative to overseas efforts. Please describe where differences in capabilities exist and what you feel are the causes for these differences.

2. If you stated that U.S. climate and weather modeling capabilities are behind those of other countries, do you have any suggestions to remedy this deficiency?

III. Computational resources

Resources currently in use

1. What are the manufacturers and models of the computer systems you rely on for your modeling efforts?

2. What was the year of installation or of the last major upgrade to each of these systems?

3. How many processors are currently operating on each of these systems?

4. What is the estimate of sustained system performance (Gflops) of each of these systems?

5. What is the central memory (GB) and secondary disc storage for each system (GB/TB)?

6. How many CPU hours are used per month? What is the cost of this time?

7. Is computational time shared with the wider community? If so, how is this interaction organized?

Future requirements needed to improve climate and weather modeling efforts

1. Please list any future upgrades that are planned to your current systems.

2. What additional upgrades would be incorporated if funds were available?

3. Does modeling capacity or capability limit your current activities or does some other factor? Could you make use of additional modeling capacity or capability for additional activities?

IV. Models

1. Please provide your thoughts on the relative merits and hindrances of running your models on massively parallel processing systems relative to parallel vector architectures.

2. Are model results produced by your facility made available to a wider scientific community? If so, are any restrictions placed on the data?

3. What climate and/or weather models are run at your institution? (please also include specifics on model type)

4. Please provide a detailed description of these models including the following:

   a. minimum grid size that can be used
   b. maximum number of vertical levels that can be used
   c. atmospheric constituents that are used to force the model including aerosols
   d. modules that are available to model earth systems (e.g. land surface vegetation, atmospheric chemistry, biogeochemistry (both land and ocean), terrestrial hydrology (both surface and soil hydrology), sea ice, etc.)
   e. treatment of boundaries in the model (e.g. swamp ocean, specifies SSTs etc.)
   f. the minimum time-step that can be used in the model

5. In an attempt to gauge the maximum achievable performance (R_{peak}) of your systems in a manner that is comparable to other systems reported in this survey we would appreciate the following information. What is the approximate run time needed to simulate 15 model years on your fastest computer using your highest resolution model? Assume that it is dedicated to the task and that optimal multi-tasking (e.g., running as many separate simulations as can be accommodated by all of the machine's processors and

dividing the final wallclock time by the number of simulations) is utilized? Please be specific as to the parameters used during this test.

6. How "portable" is your code without experiencing major performance loss?

7. What future improvements are being planned for this model?

V. Human resources

Within this section, we would like to develop a profile of the computational and human resources that are presently directed toward modeling in the U.S.

1. What is the current number of staff that are directly involved with the following and their approximate annual cost:

   a. Science/Research
   b. Hardware maintenance
   c. Software tool development
   d. Model code development
   e. Model simulation interpretation

2. Are the number of staff supporting your efforts sufficient? If not, please describe where improvements are needed.

3. When staffing positions in the categories listed above, what are the main difficulties, if any, involved (i.e. level of training required, salary requirements).

4. Please describe any future changes in staffing that are planned.

5. Are you currently planning to (or intending to in the future) convert model codes to run on massively parallel machines? If currently converting, what experience do you have with this process? If intending to in the future, what are your plans for doing so?

VI. Miscellaneous

1. Do you feel that your modeling efforts are being limited by lack of sufficient high-end computing resources? By people? By other resources? By any other factors?

2. What is your highest priority if some of these limiting factors are removed?

3. Do you have any other comments that you feel this panel should be aware of?

4. This survey was distributed to the groups listed below. If there is a group that has not been included in this list that you feel should be considered as part of our data collection, please provide us with the name and address of this institution and the person to whom this survey should be directed.

Appendix D

Questionnaire Sent to Small Modeling Centers

I. General information

1. Name of Institution:

2. Name of Research Group (if applicable):

3. Name and contact details of person completing this survey:
   Name
   Position
   Address

   Phone
   Fax
   e-mail

4. Please indicate the primary funding sources for your modeling efforts and an approximate percent breakdown where there is more than one source.

5. How would you describe the purpose(s) of your modeling efforts?

6. What percentage of your modeling activities are devoted to operational versus research purposes?

II. Interaction with major modeling centers

1. Please describe groups with which you have significant collaboration and briefly describe the nature of this collaboration.

2. If you currently have extensive collaboration with other groups, what are the main reasons for this collaboration?

3. Do you feel that your modeling effort would be aided by altering the organization of U.S. climate modeling resources? If so, what changes would you recommend be made?

III. Current U.S. modeling capabilities

1. Please provide your opinion on current U.S. climate and weather modeling capabilities relative to overseas efforts. Please describe where differences in capabilities exist and what you feel are the causes for these differences.

2. If you stated that U.S. climate and weather modeling capabilities are behind those of other countries, do you have any suggestions to remedy this deficiency?

IV. Computational resources (please answer to the best of your knowledge):

1. If you operate your own computing facilities for use in your modeling efforts, what are the manufacturers and models of these computer systems?

2. What was the year of installation or of the last major upgrade to each of these systems?

3. How many processors are currently operating on each of these systems?

4. What is the estimate of sustained system performance (Gflops) of each of these systems?

5. What is the central memory (GB) and secondary disc storage for each system (TB)?

6. How many CPU hours do you currently use per month? What is the cost of this time?

V. Future requirements

1. Please list any future upgrades that are planned to your current systems.

2. What additional upgrades would be incorporated if funds were available?

3. Does modeling capacity limit your current activities? Could you make use of additional capacity for additional activities?

VI. Models

1. What type of climate and/or weather models are run at your institution?

2. Are model results produced by your facility made available to the wider scientific community? If so, are any restrictions placed on the models or data?

3. Do you use models or outputs from other facilities? If so, are any restrictions placed on the models or data?

4. How many different models do you run? Which are the most computer intensive?

5. Please provide a brief description of these models including the following:

   a. minimum grid size that can be used
   b. maximum number of vertical levels that can be used
   c. atmospheric constituents that are used to force the model including aerosols
   d. modules that are available to model earth systems (e.g. land surface vegetation, atmospheric chemistry, biogeochemistry (both land and ocean), terrestrial hydrology (both surface and soil hydrology), sea ice, etc.)
   e. treatment of boundaries in the model (e.g. swamp ocean, specifies SSTs etc.)
   f. the minimum time-step that can be used in the model

6. In an attempt to gauge the maximum achievable performance (R_{peak}) of your systems in a manner that is comparable to other systems reported in this survey we would appreciate the following information. What is the approximate run time needed to simulate 15 model years on your fastest computer using your highest resolution model? Assume that it is dedicated to the task and that optimal multi-tasking (e.g., running as many separate simulations as can be accommodated by all of the machine's processors and dividing the final wallclock time by the number of simulations) is utilized? Please be specific as to the parameters used during this test.

7. How "portable" is your code without experiencing major performance loss?

8. What future improvements are being planned for this model?

VII. Human resources

Within this section, we would like to develop a profile of the computational and human resources that are presently directed toward modeling in the U.S.

1. What is the current number of staff that are directly involved with the following and their approximate annual cost:

    a. Science/Research
    b. Hardware maintenance
    c. Software tool development
    d. Model code development
    e. Model simulation interpretation

2. Are the number of staff supporting your efforts sufficient? If not, please describe where improvements are needed.

3. When staffing positions in the categories listed above, what are the main difficulties, if any, involved (i.e. level of training required, salary requirements).

4. Please describe any future changes in staffing that are planned.

5. Do you feel that future modeling efforts will be hindered by the availability of quality graduate students? If so, what steps would you recommend to remedy this problem?

VIII. Miscellaneous

1. Do you feel that your efforts are being limited by access to high-end computing resources? By access to model output from large modeling centers? By availability of diagnostic tools? By any other factors?

2. Do you have any other comments that you feel this panel should be aware of?

3. This survey was distributed to the groups listed below. If there is a group that has not been included in this list that you feel should be considered as part of our survey, please provide us with the name and address of this institution and the person to whom this survey should be directed.

Appendix E

Climate Modeling Survey: Summary Responses

42 Responses Received

Note: As a point of reference, there were two unique questionnaires that were sent out to U.S. modeling centers for the purposes of this report. One questionnaire was sent out to large and intermediate centers, and a second questionnaire was sent to small centers.[1] Thus, the 'coding' after each question, e.g., I6L (large/intermediate), I6S (small), specifies the question number as in the surveys above and whether it was common to both questionnaires, or exclusive to one or the other. In some instances, a question was specific to only one survey as it was believed to be inappropriate to the other category of modeling centers.

1. What percentage of your modeling activities are devoted to operational versus research purposes? (I6L, I6S)

39 Majority research oriented
3 Majority operations oriented

➢ Out of the responses that were majority research oriented, some stated that their research had direct operational relevance.

---

[1]An example of what is referred to in this document as a small modeling effort is one using a global, stand-alone atmospheric climate model at R15 (~4.5° × 7.5°) resolution; an example of an intermediate effort is one using a global, stand-alone atmospheric climate model at T42 (2.8° × 2.8°) resolution; an example of a large or high-end modeling effort is one using a global, coupled T42 atmospheric / 2° × 2° oceanic model (or finer resolution) for centennial-scale simulations of transient climate change.

2. Please describe groups with which you have significant collaboration and briefly describe the nature of this collaboration. (I7L, II1S)

➤ From the responses received, there appears to be strong connections between the major centers and academia and vice-versa.

3. Please provide your opinion on current U.S. climate and weather modeling capabilities relative to overseas efforts. Please describe where differences in capabilities exist and what you feel are the causes for these differences. (II1L, III1S)

U.S. is:

	Ahead	Behind	Comparable
Weather	2	20	6
Climate	1	21	9

Why are there differences?
Underfunded
Understaffed
Lack of computer resources
Lack of common center/coordination

Other statements:
Comparable to other countries at all but high-end
Model development is weak here and overseas
U.S. is ahead in diversity and size of effort
It is more difficult to organize the U.S. effort due to its size and diversity

4. If you stated that U.S. climate and weather modeling capabilities are behind those of other countries, do you have any suggestions to remedy this deficiency? (II2L, III2S)

7   Increased Funding
8   Shared Infrastructure
18 Enhanced Organization
25 Hardware
8   Adequate brainpower

5. Do you feel that your modeling effort would be aided by altering the organization of U.S. climate modeling resources? If so, what changes would you recommend be made? (II3S)

6   Yes
5   No

Observations of the affirmative responses:
- Too many underfunded, understaffed groups
- Inadequate links to data collection
- More emphasis should be placed on the vulnerability of the Earth system to the spectrum of environmental stresses, rather than focus primarily on the effects of greenhouse gases.
- U.S. should take the lead in the physics of the climate system and its parameterization
- Devolve computing resources away from computer centers to the users
- Develop a responsive, interactive computing environment
- Make it easier to access climate models for climate applications and to long-term model simulation data for analysis

What additional upgrades would be incorporated if funds were available? (III2bL, V2S)

7 Upgrades for PC clusters
2 More nodes
3 Increased bandwith
7 Increase general computational power
5 Increase disk storage
3 Increase file migration capabilities
1 Purchase Alpha-type workstations
1 Upgrade to parallel vector systems if possible
2 None
6 More processors

6. Does modeling capacity or capability limit your current activities or does some other factor? Could you make use of additional modeling capacity or capability for additional activities? (III3bL, V3S)

27 Yes
2   No

7   Additional human resources
18 Additional computing capabilities

7. Is computational time shared with the wider community? If so, how is this interaction organized? (III7L)

9   Yes
12 No

2 Yes, via scientific collaboration

1 Only within DOD and with DOD funded scientists
3 Sharing is through a queuing system
2 Sharing through proposals for computer time
1 Sharing only within DOE
3 Via allocation process
1 Via output only

8. Please provide your thoughts on the relative merits and hindrances of running your models on massively parallel processing systems relative to parallel vector architectures. (IV1L)

4   Massively parallel architecture is better
18   Parallel vector architecture is better

MPP architecture is better but:
– There are a lack of compilers for these systems
– The transfer of code to MPP is not easy
– Vendors are not ready to supply the needed systems

Other comments:
MPP is harder to use
MPP benefit is that the processing time is cheaper as the cost of the systems and maintenance is less than for parallel vector systems
MPP offers more CPU power and memory per dollar spent
Some new models can only be run on MPP
MPP requires longer code development
MPP is not scalable
MPP offers poor system software and is unstable
MPP requires additional personnel
MPP offers poor communication among processors

9. Do you use models or outputs from other facilities? If so, are any restrictions placed on the models or data? (VI3S)

12   Yes
0   No

Restrictions:
Output is restricted to research collaborators
DOE security restrictions on computing access
No restrictions
Some foreign data is restricted
Some data are restricted due to being in a pre-release state

10. How "portable" is your code without experiencing major performance loss? (IV6L)

22 Very portable
2   Somewhat portable
8   Code is custom altered for specific platforms
1   Code works on MPP only
1   Code is portable to VPP and MPP with some limitation

11. Are you currently planning to (or intending to in the future) convert model codes to run on massively parallel machines? If currently converting, what experience do you have with this process? If intending to in the future, what are your plans for doing so? (V5L)

10 Already converted
12 Underway
2   Not underway

12. Are model results produced by your facility made available to the wider scientific community? If so, are any restrictions placed on the models or data? (VI2S)

10 Yes
0   No

Are model results produced by your facility made available to a wider scientific community? If so, are any restrictions placed on the data? (IV2L)

3   Yes with some restrictions
21 Yes
1   Yes, but only with collaborators
0   No

Additional:
2   More widely distributed if resources were available
1   Yes, through published work

13. Are the number of staff supporting your efforts sufficient? If not, please describe where improvements are needed. (V2L)

6   Yes
20 No
Staff needed for:

Data interpretation and analysis
Programmers
Software engineers
Hardware maintenance
Model simulation interpretation
High-performance applications

Are the number of staff supporting your efforts sufficient? If not, please describe where improvements are needed. (VII2S)

4 Yes
8 No

Staff needed for:
Data interpretation and analysis
Programmers
Model developers

14. Do you feel that your efforts are being limited by access to high-end computing resources? By access to model output from large modeling centers? By availability of diagnostic tools? By any other factors? (VIII1S)

11 Yes
1   No

1 Skilled personnel are not centrally located
1 No long-term strategy
1 Data outputs need to be made more user friendly
1 Satellite data needs to be made more user friendly
9 Access to computing
1 Access to global models
1 Stable funding

Do you feel that your modeling efforts are being limited by lack of sufficient high-end computing resources? By people? By other resources? By any other factors? (VI1L)

26 Yes
27 No

Factors:
17 People
18 Computing

Other factors:

Lack of well-documented modern model codes
Network bandwidth
Data storage
Stable funding

15. When staffing positions in the categories listed above, what are the main difficulties, if any, involved (i.e. level of training required, salary requirements). (V3L, VII3S)

1 Research is very specialized
16 Salary is not competitive
10 Finding funding
15 Level of training
7 Difficult to find qualified programmers
1 Navy bureaucracy
3 Difficult to find model developers
1 No difficulty

16. Please describe any future changes in staffing that are planned. (V4L)

7 None
8 Model/software support
8 Scientist
5 Modeler
1 Hardware

Please describe any future changes in staffing that are planned. (VI4S)

4 None
4 Model/software support
3 Scientist
0 Hardward maintenance

17. What is your highest priority if some of these limiting factors are removed? (VI2L)

11 Enhanced computing capabilities
8 Enhanced human resources
7 Improved physical performance of the models
1 Build a modeling system infrastructure
4 Increase the number of models
7 Increase model resolution

1 Develop a high performance regional climate model
1 Adapt model code for parallel systems
1 Perform simulations on non-local systems
1 Additional R& D research funding

18. Do you feel that future modeling efforts will be hindered by the availability of quality graduate students? If so, what steps would you recommend to remedy this problem? (VI5S)

3 No
5 Yes

Appendix F

Workshop Agenda

IMPROVING THE EFFECTIVENESS OF CLIMATE MODELING
AUGUST 21–23, 2000
NATIONAL RESEARCH COUNCIL
WASHINGTON, D.C.

AGENDA
Monday, August 21ˢᵗ   Lecture Room

OPEN SESSION

8:00 a.m.	Breakfast	
8:45 a.m.	Welcome and Introduction	Joe Friday
9:00 a.m.	Large-Scale Modeling in the United States	Maurice Blackmon
9:45 a.m.	Large-Scale Modeling in Europe	Lennart Bengtsson
10:30 a.m.	Parallel Supercomputing for Weather and Climate	Matthew O'Keefe
11:15 a.m.	Curiosity-Driven vs. Product-Driven Research	Ricky Rood

12:00 p.m.   Paths Towards Increasing the
                Effectiveness of U.S. Climate   Ed Sarachik
                Modeling

12:30 p.m.   Working Lunch

1:30 p.m.   Afternoon Session: Interactive discussion between the
                attendees and the panel

        Topics:
        — How can we constitute large-scale modeling in response to
        national needs?
        — How can we increase the benefits to the research community
        responding to national needs?

5:00 p.m.   Adjourn for the day and dinner in the Members Room

Appendix G

Workshop Participants

Phil Arkin	National Oceanic and Atmospheric Administration
Bob Atlas	National Aeronautics and Space Administration
Dave Bader	Department of Energy
Anjuli Bamzai	National Science Foundation
Lennart Bengtsson	Max Planck Institut fur Meteorologie
Alan Betts	Atmospheric Research
Maurice Blackmon	National Center for Atmospheric Research
Jay Fein	National Science Foundation
Carter Ford	National Research Council
Joe Friday	National Research Council
W. Lawrence Gates	Lawrence Livermore National Laboratory
Bryan Hannegan	Senate Energy and Natural Resources Committee
Timothy Hogan	Naval Research Laboratory
Alexandra Isern	National Research Council
Tim Killeen	National Center for Atmospheric Research
Ants Leetma	National Oceanic and Atmospheric Administration
Margaret Leinen	National Science Foundation
Margaret Lemone	National Center for Atmospheric Research
S.-J. Lin	National Aeronautics and Space Administration
Eric Lindstrom	National Aeronautics and Space Administration

Jerry Mahlman	National Oceanic and Atmospheric Administration
Robert Malone	Los Alamos National Laboratory
John Marshall	Massachusetts Institute of Technology
Roberto Mechoso	University of California-Los Angeles
Chris Miller	National Oceanic and Atmospheric Administration
Kenneth Mooney	National Oceanic and Atmospheric Administration
Matthew O'Keefe	University of Minnesota
David Randall	Colorado State University
Michele Rienecker	National Aeronautics and Space Administration
Richard Rood	National Aeronautics and Space Administration
Edward Sarachik	University of Washington
Albert Semtner	Naval Postgraduate School
Peter Schultz	National Research Council
Max Suarez	National Aeronautics and Space Administration
Jagadish Shukla	George Mason Univ. Center for Ocean-Land-Atmosphere
Ronald Stouffer	National Oceanic and Atmospheric Administration
James Todd	National Oceanic and Atmospheric Administration
Stephen Zebiak	International Research Institute for Climate Prediction

Appendix H

Summary of Other Relevant Reports

i. The Report of the DOE/NSF National Workshop on Advanced Scientific Computing.

This workshop report (commonly called the "Langer Report") was designed to provide a foundation for establishing a science-driven national infrastructure of terascale computing, communications, and advanced simulation. Weather and climate prediction were given as examples of scientific and engineering applications needing such expanded computing facilities. This report discussed the following obstacles to the successful use of terascale computing facilities: (a) the scarcity of human resources applied to computational and scientific research problems; (b) the difficulty of matching the financial rewards offered by private industry, and; (c) the lack of proper software available to usefully optimize performance on the new generations of massively parallel computers.

Relevant recommendations arising from this workshop were:

1. the U.S. should launch a vigorous effort to make high-speed computing systems accessible to the national scientific and engineering communities;

2. the U.S. should concurrently launch a vigorous effort to develop software, algorithms, communication infrastructure, and the visualization systems necessary for effective use of the next generation of computing facilities;

3. the U.S. scientific and engineering communities should prepare to use these computing facilities to solve complex problems of both basic and strategic importance;

4. hardware, software and communications developments should be coordinated with these scientific and engineering applications (DOE/NSF, 1998).

ii. Report of the NSF/NCEP Workshop on Global Weather and Climate Modeling

This interagency workshop report discusses the future of weather and climate modeling in the United States. An important outcome of this report was the recognition that the diversity of U.S. modeling has created a barrier to efficient collaboration between various modeling groups. Recommendations were:

1. a common modeling infrastructure should be established to facilitate the evaluation and exchange of technological and research advances in the broader modeling community, with exchanges envisioned not only between the operational and research communities but also between the numerical weather prediction and climate modeling communities;

2. the common modeling infrastructure should be advanced by establishing modeling standards and guiding principles and by focusing efforts on the development of a finite number of core models, each of which would be devoted to a major area of modeling (e.g., numerical weather prediction, seasonal to interannual prediction, decadal variability);

3. the National Centers for Environmental Prediction be one of the centers associated with a core model promoting the common modeling infrastructure because of its responsibility for U.S. operational forecasts and because of its critical data assimilation activities.

iii. President's Information Technology Advisory Committee

The President's Information Technology Advisory Committee on future directions in information technology (PITAC, 1999) recommended increased investment in information technology, with priorities in the areas of software, information infrastructure (including networks), and high-end computing systems, noting that "extremely fast computing systems, with both rapid calculation and rapid data movement, are essential to provide accurate weather and climate forecasting ... to conduct scientific research in a variety of different areas and to support critical national interests."

iv. Accelerated Climate Prediction Initiative

An interagency committee charged with making recommendations on the implementation of the Department of Energy Accelerated Climate Prediction Initiative (ACPI; Gates et al., 1999) produced a report that argued for a climate modeling structure that would include a centralized computing facility; a modeling and research consortium of exclusive users of the facility for specific projects involving model development, model diagnoses and predictions on all time scales; and a group of regional climate centers (recommended to be initially three to five in number) that would interface with local user communities to examine the impacts of climate change and variability on regional scales. The committee also recommended a national software and communications infrastructure to "provide a significantly enhanced capability to store, access, transfer, diagnose and visualize the results of high-end climate model simulations, [and to] provide an effective software and communications network linking the project's components and participants ..."

A review of the ACPI program by the JASON Group noted that (JASON, 1998):

1. substantial increases in the computational power available to U.S. researchers are well warranted and can contribute to a better understanding of the climate system;

2. computational power alone will not greatly improve our abilities to predict climate, and linked observational programs and process studies are also essential for a balanced global change effort.

v. High-End Climate Science: Development of Modeling and Related Computing Capabilities (USGCRP, 2000).

This report, solicited by the Office of Science and Technology Policy, was designed to make recommendations that "responds to unmet national needs in climate prediction, climate-science research and climate-change assessment." The basic recommendation of this report is the creation of a dedicated organization, called the "Climate Service," designed to respond to the various requirements and demands placed on the climate community. The report distinguishes between traditional research (discovery-driven) and the type of activity required by new demands on the climate community (product-driven). This Climate Service should be product-driven, centralized, and requires a new business plan to manage the Service in an integrated manner.

The report recommends that the Climate Service should have access to computational systems with the highest level of capability and should engage in two major core simulation activities: one in weather and one in climate. It noted that the obstacles to forming the Service are organizational and computational. In particular, it noted that the U.S. policy on high performance computing, imposing restrictive duties on Japanese vector supercomputers, significantly complicates the people problem for the Climate Service in that massively parallel distributed-memory computers requires a much greater investment in software, both system and applications, and therefore requires just the type of information technologists in greatest demand in a booming software industrial environment.

In order to remedy the fragmentation of effort discussed in NRC, 1999a, the report also recommends the Climate Service seek integration of effort across the disciplines involved in weather and climate modeling, across institutions, and across modeling, data and computational systems. A software infrastructure to enable collaboration and development of software by multiple scientists at multiple institutions, to allow a transition path from discovery-driven research to product-driven activities, and to optimize across computational resources and scientific research is also recommended. Issues of machines and software form an important part of the report.

Appendix I

Description of Different Codes

a. ECMWF's IFS (Integrated Forecast System)

The ECMWF's IFS code is a parallel spectral weather model that is also used for seasonal climate prediction. Its structure is similar to climate codes from NCAR, including CCM, but its parallel execution model is highly evolved. It uses domain decomposition in two-dimensions and performs both spectral and Legendre transformations on the grid data. The sustained rates reported for IFS are in units of forecast days per day (in other words, the ratio of simulated time to real wall-clock time). The two machines compared with this code are the T3E (600-MHz Alpha processors) and the Fujitsu VPP5000. A single processor of the VPP5000 achieves 48 times the sustained speed of a single Cray T3E processor. In parallel configurations this ratio (of sustained rates per processor) increases to 57 when IFS executes on 1408 T3E processors and 98 VPP5000 processors. Notice that only 98 VPP5000 processors are nearly four times faster than 1,408 T3E processors.

Analysis of the performance of the IFS code (at T213 L31 resolution) on a variety of machines, (both microprocessor- and vector-based) indicates that machines with small numbers of fast vector processors are superior to highly parallel microprocessor-based SMPs (Fig. A-1).

b. Environment Canada's MC2

MC2 is a regional, non-hydrostatic weather model. It uses a variety of sophisticated solvers and is structured somewhat like a global spectral model. In Table 4-3 we show the sustained performance in Mflops of MC2 for the Origin 2000 (250-MHz R10000 processors with 4-MB caches) ver-

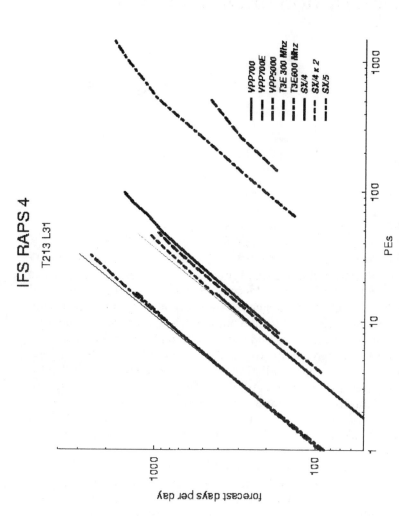

FIGURE A-1 Performance of the IFS code (at T213 L31 resolution) on a variety of machines, both microprocessor- and vector-based. The figure shows that machines with small numbers of fast vector processors are superior to highly parallel microprocessor-based SMPs. Courtesy of D. Dent, European Centre for Medium-range Weather Forecasts.

sus the NEC SX-5, a Japanese VPP. Like the IFS code, the MC2 code executes nearly 50 times faster (on a per-processor basis) on the SX-5 compared to the R10000. The aggregate performance achieved on 28 SX-5 processors is 95,200 Mflops, almost 50% of the potential peak speed of the SX-5 in this parallel configuration. These speeds dwarf the 840 Mflops achieved on a smaller 12-processor configuration of the Origin.

c. NCAR's MM5

MM5 is a grid-based, hydrostatic mesoscale model originally developed at Penn State. Its performance on a single Alpha 667-MHz processor is 360 Mflops, about 27% of peak performance. This is a much higher percentage of the peak microprocessor speed than is achieved by most weather models and is due primarily to MM5's good cache useage. When parallel execution of MM5 is considered, the AlphaServer cluster (128 4-processor machines connected together into a 512-processor configuration) is about 60% faster than a 20-processor VPP5000.

The per-processor speed ratio between the 512-processor Alpha and 20 processor VPP5000 is 16 in favor of the vector machine, much less than the 50-times difference found with the other codes. However, the lower ratio is consistent with the MM5, which attains sustained-to-peak performance a factor of 3 higher than the other codes.

d. German Weather Service's LM and GME

Only limited performance data was available for the local (LM) and global (GME) weather models used by the German Weather Service (Deutscher Wetterdienst). LM is a regional model, while GME is a global model developed using an icosahedral-hexagonal grid. Both these models execute nearly 50 times faster (either serial or parallel) on a VPP5000 than on a Cray T3E.

The actual performance of real weather and climate codes support our contention that currently Japanese parallel vector supercomputers significantly outperform American-manufactured MPPs based upon microprocessor technology. If VPPs are not available, it is more difficult to get good performance. This is particularly true for "capability computing" (see Section 3-1). As was demonstrated above, Amdahl's law requires a very high degree of parallelism in a model to achieve effective speedups on large numbers of processors. Recognizing this, SMP vendors have been moving to a hybrid architecture that places multiple processors on each node. SMP clusters typically have nodes that contain 2–16 processors sharing uniform memory access (UMA) via a bus or, when the number of processors exceeds 8, a higher-performance intra-node network. To take advantage of this UMA feature the preferred intra-node programming model is OpenMP threads, an evolving standard for what was once known on CRI VPP machines as "multi-tasking." MPI is used for the non-uniform memory access (NUMA) inter-node communications. Although this hybrid programming model adds another layer of complexity, it offers a useful path to parallelism if it can be efficiently implemented. For example, rather than doing two-dimensional domain decomposition in longitude and latitude, with the hybrid model one might decompose and use MPI only in latitude while treating parallelism in longitude with threads spread across the processors on each node.

Important considerations from a software standpoint are

1. optimizing the placement of data with respect to the processor(s) that will use it most;

2. minimizing the number and maximizing the size of messages sent between nodes;

3. maximizing the number of operations performed on data that is in cache, while minimizing the amount of data required to be in cache for these operations to occur.

Normally, one expects the operating system or job scheduler to take care of "1" automatically. If data is not localized on the same node as the processor that will use it most often, performance will suffer and is likely to be quite variable from run to run. Item "2" requires careful planning of MPI calls. Item "3" requires the most code changes, such as subdividing the computational domain into blocks small enough so that all data for any single block will fit into cache. More radical steps involve converting from Fortran 90 to Fortran 77 in order to get explicit do-loops, re-ordering array and loop indices, in-lining subroutine calls, fusing loops, and other optimizations that would be left to the compilers if only they were capable.

Appendix J

Acronyms

ACPI	Accelerated Climate Prediction Initiative
AGU	American Geophysical Union
AMS	American Meteorological Society
AO	Arctic Oscillation
ARM	Atmospheric Radiation Measurement
ARPS	Advanced Regional Prediction System
ASP	Application Service Provider
CCN	Cloud Condensation Nuclei
CCPP	Climate Change Prediction Program
CDC	Climate Diagnostics Center
CFC	Chlorofluorocarbon
CIT	California Institute of Technology
CMI	Common Modeling Infrastructure
CMIP	Coupled Model Intercomparison Project
COAMPS	Coupled Ocean/Atmosphere Mesoscale Prediction System
COARE	Coupled Ocean Atmosphere Response Experiment
COLA	Center for Ocean-Land-Atmosphere Studies
CONV	Conventional Data
CPU	Central Processing Unit
CRC	Climate Research Committee
CRI	Cray Research, Inc.
CRM	Cloud-Resolving Mode
CSM	Climate System Modeling
CSU	Colorado State University

DAO	Data Assimilation Office
DAS	Data Assimilation System
DEC	Digital Equipment Corporation
DOE	U.S. Department of Energy
DSP	Dynamical Seasonal Prediction
ECMWF	European Centre for Medium-range Weather Forecasts
ENSO	El Niño/Southern Oscillation
EOS	Earth Observing System
EPA	Environmental Protection Agency
EUMETSAT	European Meteorological Satellite
FGGE	First GARP Global Experiment
FIFE	First International Satellite Land Surface Climatology
FLOPS	Floating Point Operations per Second
FSU	Florida State University
GARP	Global Atmospheric Research Programme
GATE	GARP Atlantic Tropical Experiment
GCIP	GEWEX Continental-Scale International Project
GCM	General Circulation Model
GCSS	GEWEX Cloud System Study
GEOS	Goddard Earth Observing Satellite
GEWEX	Global Energy and Water Cycle Experiment
GFLOP	GigaFlop
GFDL	Geophysical Fluid Dynamics Laboratory
GISS	Goddard Institute for Space Studies
GLA	Goddard Laboratory for Atmospheres
GSFC	Goddard Space Flight Center
HPCC	High Performance Computing and Communications
IBM	International Business Machine
IFS	Integrated Forecast System
INDOEX	Indian Ocean Experiment
IPCC	Intergovernmental Panel on Climate Change
IRI	International Research Institute
ISLSCP	International Satellite Land Surface Climatology Project
IT	Information Technology
ITCZ	Intertropical Convergence Zone
JPL	Jet Propulsion Laboratory
LANL	Los Alamos National Laboratory
LAWS	Laser Atmospheric Wind Sounder
LES	Large Eddy Simulations
MFLOP	MegaFlop
MOS	Model Output Statistics
MPI	Message Passing Interface

MPP	Massively Parallel Processor
NAO	North American Oscillation
NAS	National Aerodynamic Simulation
NASA	National Aeronautics and Space Administration
NCAR	National Center for Atmospheric Research
NCEP	National Centers for Environmental Prediction
NERSC	National Energy Research Scientific-Computing Center
NOAA	National Oceanic and Atmospheric Administration
NPGS	Naval Postgraduate School
NRC	National Research Council
NRL	Naval Research Laboratory
NSCAT	NASA Scatterometer
NSF	National Science Foundation
NUMA	Non-Uniform Memory Access
NWP	Numerical Weather Prediction
OLR	Outgoing Longwave Radiative Flux
OSE	Observing System Experiments
OSSE	Observing System Simulation Experiments
PC	Personal Computer
PDO	Pacific Decadal Oscillation
PE	Processing Element
PI	Principal Investigator
PILPS	Project for Intercomparison of Landsurface Parameterization Schemes
PITAC	President's Information Technology Advisory Committee
PNNL	Pacific Northwest National Laboratory
PROVOST	Prediction of Climate Variations on Season-to-Interannual Time Scales
PSU	Pennsylvania State University
RAMS	Regional Atmospheric Modeling System
RMS	Root Mean Square
ROC	Relative Operating Characteristic
SATEMS	Satellite Temperature Soundings
SCM	Single-Column Model
SGI	Silicon Graphics International
SMP	Shared-Memory Processor
SMP	Symmetric Multi-Processor
SSM/I	Special Sensor Microwave/Imager
SST	Sea Surface Temperature
SV	Scalable Vector
TAO	Tropical Atmosphere Ocean array
TFLOP	Teraflop

TIROS	Television Infrared Operational Satellite
TIROS-N	Television Infrared Operational Satellite - Next-generation
TOGA	Tropical Ocean-Global Atmosphere
UCAR	University Corporation for Atmospheric Research
UCLA	University of California at Los Angeles
UH	University of Hawaii
UI	University of Illinois
UMA	Uniform Memory Access
UNEP	United Nations Environment Program
USGCRP	United States Global Change Research Program
USWRP	United States Weather Research Program
VPP	Vector Parallel Processor
WCRP	World Climate Research Programme
WMO	World Meteorological Organization
WRF	Weather Research and Forecasting
XBT	Expendable Bathy-Thermographs